Big Data

STUDIES IN POLITICS, SECURITY AND SOCIETY

Edited by Stanisław Sulowski
Faculty of Political Science and International Studies
University of Warsaw

VOLUME 60

Aleksander Żołnierski / Dariusz Jaruga (eds.)

Big Data

Post-Truth and Disinformation in Politics and Economy

PETER LANG

Berlin - Bruxelles - Chennai - Lausanne - New York - Oxford

Bibliographic Information published by the Deutsche Nationalbibliothek
The Deutsche Nationalbibliothek lists this publication in the Deutsche Nationalbibliografie; detailed bibliographic data is available in the internet at http://dnb.d-nb.de.

Library of Congress Cataloging-in-Publication Data
A CIP catalog record for this book has been applied for at the Library of Congress.

Aleksander Żołnierski - Institute of Economics, Polish Academy of Sciences

Dariusz Jaruga - Faculty of Political Science and International Studies, University of Warsaw

The publication leading to the described results was carried out as part of the project “Identification, Collecting and evaluation of hostile disinformation operations in cyberspace (ICON)” co-financed by The National Centre for Research and Development within the Scientific Research and Development Program “CyberSecIdent - Cybersecurity and e-Identity”. Beneficiaries: Centre for Refining Information and the University of Warsaw. Total value of the project: 8.230.856 PLN; co-financing value: 6.605.669 PLN. Agreement number: CYBERSECIDENT/489281/IV/NCBR/2021. Completion period: 01/09/2021 – 31/03/2025.

The cover image courtesy of Benjamin Ben Chaim.

ISSN 2199-028X
ISBN 978-3-631-92243-9 (Print)
E-ISBN 978-3-631-92760-1 (E-Book)
E-ISBN 978-3-631-92761-8 (E-PUB)
DOI 10.3726/b22416

Published by Peter Lang GmbH, Berlin, Deutschland

info@peterlang.com - www.peterlang.com

This publication has been peer reviewed.
www.peterlang.com

Contents

Aleksander Żołnierski
Introduction 7

Wiesław Cetera, Dariusz Jaruga, Piotr Celiński and Jan Grzegorek
I Lie, Therefore I Am 11

Katarzyna Materska
Contributions of Polish Projects to Countering Disinformation (2019–2022) 29

Aleksander Żołnierski and Jan Grzegorek
Migrants: Reactions to the Refugee Crisis 2021–2022 47

Iuliia Arkhipova
Dissemination of Russian Propaganda via Social Media: The Case of the Vinnytsia Missile Strikes 67

Halyna Synorub
Ways of Forming Informational Resilience of the Audience Under Wartime Conditions 75

Olesia Medynska
Disinformation Narratives of Social Media in the Context of Social Transformations 97

Jakub Kruk
Political Narratives in Media: Analysis of Sentiments in Statements of Politicians and Their Impact on the Public Opinion 117

Magdalena Krutul
Emotions on News Websites During Global Crises: Analysis of Sentiment in Journalistic Publications 149

Bogusław Jagusiak and Patryk Szczepkowski
The Significance of Social Media in the Disinformation Process 179

Aleksander Żołnierski

Introduction

What we offer to readers here is the next publication on big data and the use of artificial intelligence in the practice of science, economy and politics. This time our monograph focuses primarily on security matters, issues that are demanded by the international political and economic situation. Disinformation is a tool of hybrid war, shaping the social and economic environment long before the start of the Russian invasion of Ukraine. Since February 2022, the issues of post-truth, fake news or disinformation have taken on an additional gravity.

The monograph is a result of the collaboration of scholars and experts from many academic centres and organisations, both domestic and international. The richness of the subjects addressed was achieved by inviting specialists from the University of Warsaw, the Institute of Economics of the Polish Academy of Sciences, the University of Silesia, the Gdańsk Medical University, the Faculty of Cybernetics of the Military Academy of Technology, the University of Technology and Economics in Warsaw, the Karkonosze University of Applied Sciences and the Ternopil Volodymyr Hnatiuk National Pedagogical University (Ukraine) to participate in the project.

In recent years, the application of artificial intelligence and big data analysis has become increasingly common, also in areas related to security and cybersecurity. AI and BD contribute to the increased quality of life and security of societies and individuals.

The use of software based on the advanced algorithms of artificial intelligence and advanced utilities of big data analysis enabled a new critical perspective on many different areas of the world around us. The purpose of this volume is to present the perspectives of researchers, experts and scientists, who care about the problem of distortion of the image of reality by the authors of false narratives.

"I lie, therefore I am" – these words are used to start the narrative by Wiesław Cetera and his team, for which the issues of disinformation, post-truth, use of lies and manipulations in political activities were the central problem in the context of the war beyond our eastern border. The issues of managing false information have become crucial for the various processes occurring in parallel with the armed conflict in Ukraine. Fake news and post-truth relate to direct military operations, but also to the operations of the cyber-war and hybrid war that Russia declared on the democratic world. Manipulating information also constitutes an important part of the political landscape of Poland in the pre-election period and the impact of operations shaping the energy policy that are related to or derivative of the armed conflict in Europe is felt far beyond the borders of the continent itself.

In the following chapter, Katarzyna Materska presents the contributions of Polish research projects to countering disinformation. She addresses disinformation understood in qualitative terms as the capacity to create a manipulated image of reality in a specific purpose. The environment, in which the fake news factories are founded that forge weapons in the service of manipulation, requires systematic and careful monitoring, analysis and creation of diverse and courageous scenarios of counteracting falsehoods.

Aleksander Żołnierski and Jan Grzegorek address the problems of reactions to the humanitarian crisis on the Polish-Belarusian border. A wave of racism flooded the Polish Internet and the factor differentiating the approach to these migrants was their skin colour. Although the humanitarian crisis was undoubtedly planned in Moscow, Polish security services played an essential role in it. The question is whether the engagement of security services could have indicated that this was a dress rehearsal before the invasion on Ukraine and if the anti-immigrant attitude of Poles would benefit Moscow?

Jakub Kruk, in his analysis of political narratives in the media, based on the analysis of sentiments, describes the impact of the statements of politicians on public opinion. The author finds, however, a correlation between the support for the individual political formations and the evaluation of the current political situation in Poland. The immense disproportion in the evaluation of the current situation points to the existence of the information bubble created by politicians and media connected with the Law and Justice (PiS) political party.

Magdalena Krutul, in her description of emotions on news websites during global crises, formulates the hypothesis that during media crises the quantity of content with negative overtones increases. Based on the analysis of sentiments, in the period of the pandemic, relatively more content with positive overtones appeared on news websites, but much more negative content appeared in the initial period of the war in Ukraine. In view of such research results, the hypothesis cannot be unambiguously confirmed.

The issues of the meaning of social media in the disinformation process are analysed by Bogusław Jagusiak and Patryk Szczepkowski. They conclude their analysis by stating that in a world of electronic media, society is exceptionally vulnerable to manipulation. In this process, social media are not only tools for rapid propagation of information but also an efficient means of disinformation.

The dissemination of false information can be presented as a process of diffusion of innovations, of which information is the product.

Iuliia Arkhipova investigates the phenomenon of the dissemination of Russian propaganda on social media. The dissemination of Russian propaganda is a complex process involving different entities from state-sponsored media, such as RT, to regular bloggers. These could be war correspondents, aspiring to be opinion leaders and starting a political career. Other entities include anonymous channels aiming to obtain a large number of subscribers and start the sale of advertisements.

Halyna Synorub, in her text on the methods of shaping information resilience, describes the changes in the ways in which the media operate under conditions of war. She emphasises the importance of critical thinking as the desirable behavioural strategy of the media audience. The author analyses the strategies of informative and psychological operations that consist in demoralising the population, weakening the support for the army and breakdown of Ukrainian society. The article also described the importance of emotional stability for building information resilience.

Olesia Medynska addresses the subject of disinformation narratives in social media. In the identification of the main approaches to the classification of disinformation in academic and practical media discourse, the author presents the results of social media monitoring to identify structures of disinformation. In the article, the author presents the most frequent disinformation narratives and explains their destructive impact not only on the decision-making system but also on the social awareness, social assessment and the opinions of citizens.

We hope that the richness of content and the multitude of interesting and often new research areas presented will inspire further scholarly research and further effective implementation of tools based on big data methods.

Wiesław Cetera, Dariusz Jaruga, Piotr Celiński
and Jan Grzegorek

I Lie, Therefore I Am

Internet statistics suggest that we have always been lying, sometimes less and sometimes more. Since the year 2000, interest in the term "fake news" has been increasing exponentially (Fig. 1). New information distribution channels, including social media, have created new possibilities for the dissemination of information, which increased the "efficiency" of the lie. At present, this is a new weapon in the hands of enemies.

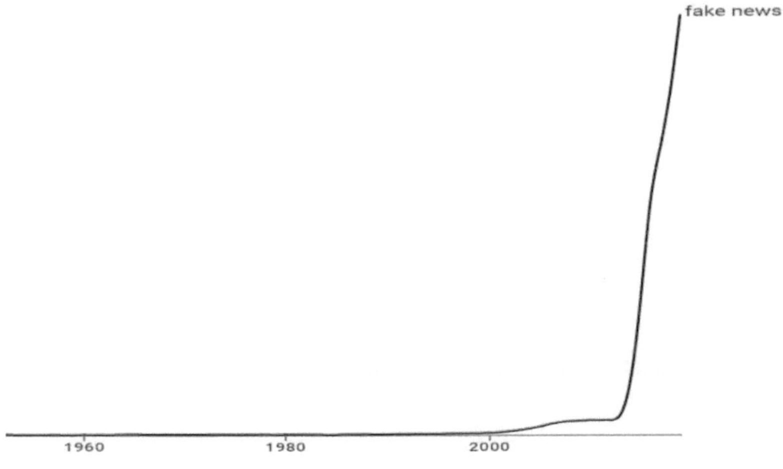

Fig. 1. Increase in interest in the term "fake news" (*Source:* ngrams)

Based on the research conducted by ChronPESEL.pl and Krajowy Rejestr Długów under the auspices of the Personal Data Protection Office, only 21.6 % of respondents are completely certain that they could recognise false news reports.[1] In the 35–54 age group, this is only ca. 16 % of respondents, and among those over 65, only 12 %. Yet we live in an environment that is hostile

1 https://chronpesel.pl/aktualnosci/mlodzi-polacy-nie-dbaja-o-swoje-hasla-do-logowania (accessed: 12.12.2022).

to the truth. There are ongoing wide-scale disinformation operations initiated by Russia that the Centre for Eastern Studies described as the Russian information war.[2]

Telegram

Telegram is an instant messenger that is popular in Russia and connected with a network of public pages. On this forum, news reports are discussed by a network of government-connected bloggers (including the television "talking heads") and anonymous public pages. Most politically active Russian celebrities select this instant messenger as the primary platform for communication with their audience. The news reports from bloggers (including those working in the war zone) and anonymous channels are re-posted by celebrities. Telegram is rarely censored and vulgar language and dehumanising narratives are commonly used.

For the analytical purposes, the author collected 1,665 posts published by Russian propaganda channels on Telegram in the period from 1 to 15 December 2022. The posts collected reached 148M impressions in total, including approx. 26.4M impressions of reposts from sources not included in the database. The initial database consisted of 342 Telegram channels and the posts published by 207 of them were included in the posts database (the remainder had no related posts). Propaganda channels were pre-identified using the "snowball" method and their activity started from the accounts of people connected with Russia Today (RT). The research did not include the posts published by channels closely connected with RT (e.g. Sputnik), as these are blocked in Poland. The search was conducted using the word "Польша"/ "Polska" (Poland) (by default, the search results on Telegram include words with different suffixes, such as "Польши", "Польше", "Польшу", but not "polski" (Polish) or "Polak" (Pole) – "польский", "поляк").

The number of posts, in which Poland is mentioned, depended on the events taking place on a given day (Fig. 2). For example, the highest number of posts on 15 December 2022 was caused by the approval by the European Union parliament of the EUR 18B support package for Ukraine for 2023, and… the explosion in the headquarters of the Polish Police (the police chief fired a grenade launcher that he had brought as a gift from his visit to Ukraine) and the reaction to the identification of Russia as a state sponsor of terrorism.

2 K. Chawryło Propaganda masowego rażenia. Rosyjska telewizja w obliczu wojny. Comments of the Centre for Eastern Studies. No. 443. 06.05.2022 (accessed: 12.12.2022).

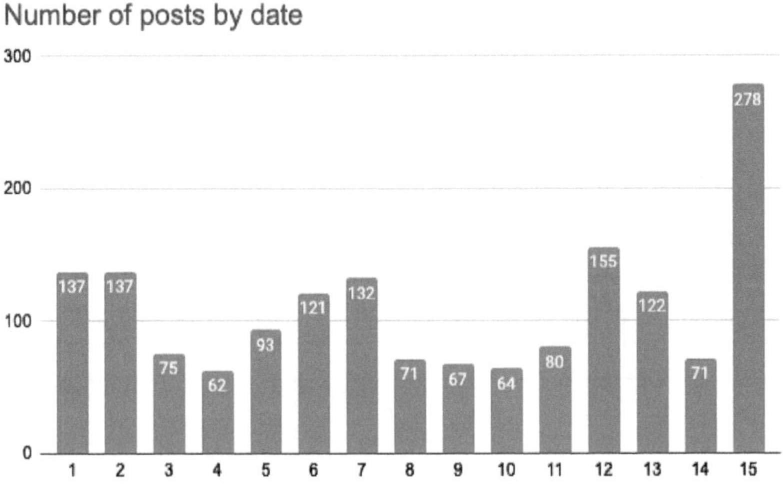

Fig. 2. Number of posts on Poland on Telegram instant messenger pages (*Source:* original research)

The Telegram channels with the greatest reach may be divided into state-owned media (rian_ru, ukraina_ru), private media (readovkanews, readovkaru, stranaua, breakingmash), celebrities (yurasumy, SolovievLive, boris_rozhin, ASupersharij, dimsmirnov175, VA_Nikonov) and anonymous channels (Tab. 1).

Tab. 1. Telegram channels with the highest range

Name	Number of views	Number of posts	Average views per post
rian_ru	13,289,100.00	19	699,426.32
readovkanews	12,961,700.00	23	563,552.17
yurasumy	12,225,400.00	7	1,746,485.71
SolovievLive	11,544,900.00	39	296,023.08
bbbreaking	7,564,100.00	15	504,273.33
boris_rozhin	5,987,800.00	11	544,345.45
rybar	4,245,100.00	11	385,918.18
kremlinprachka	3,984,500.00	11	362,227.27
RVvoenkor	3,838,000.00	9	426,444.44
readovkaru	3,013,900.00	10	301,390.00
ASupersharij	2,919,900.00	4	729,975.00

(*continued on next page*)

Tab. 1. Continued

Name	Number of views	Number of posts	Average views per post
ukraina_ru	2,875,100.00	52	55,290.38
dimsmirnov175	2,542,600.00	14	181,614.29
stranaua	2,456,600.00	35	70,188.57
breakingmash	2,189,100.00	3	729,700.00
voenacher	2,169,600.00	11	197,236.36
swodki	2,065,900.00	53	38,979.25
ssigny	2,050,100.00	21	97,623.81
brozabrata	2,034,100.00	12	169,508.33
VA_Nikonov	1,879,600.00	8	234,950.00

(*Source:* original research)

As far as the number of posts per channel, there is no dominant source specialised in Polish issues among the sources included in the database (Tab. 2). However, the most active channel that published 69 posts including the word "Polska" (Poland) was Belarus_VPO, which has declared that it is managed by Belarusian soldiers. The Swodki channel, which published 53 posts, usually republishes the texts written by other authors on different platforms.

Tab. 2. Telegram channels with the most posts on Poland

Name	Number of posts	Average views per post	Number of views
Belarus_ VPO	69	6,802.90	469,400.00
swodki	53	38,979.25	2,065,900.00
ukraina_ru	52	55,290.38	2,875,100.00
tass_agency	42	36,035.71	1,513,500.00
baltnews	42	3,673.81	154,300.00
SolovievLive	39	296,023.08	11,544,900.00
belvestnik	39	5,502.56	214,600.00

(*Source:* own research)

Two channels that specialise in Polish issues were found, because their publications were re-posted on the channels included in the initial database: PL_Syrenka (ca. 20,000 followers) and PanSwiridow (ca. 3,500 followers). PL_Syrenka is an anonymous channel, but based on the excessive number of reposts from and by

BaltNews, it may be concluded that they are connected and BaltNews is part of Russia Today media holdings. Almost every post by PL_Syrenki is undersigned with the saying "Польша – не заграница" (part of the Soviet saying "Курица не птица, Польша / Болгария / Украина – не заграница" / "A chicken is not a bird, Poland / Bulgaria / Ukraine is not a foreign country"). The PanSwiridow channel is administered by Leonid Sviridov, who works for Sputnik / RT. The subjects and narratives of these channels do not significantly deviate from the narratives of the mainstream channels.

The Polish politician mentioned the most by Russian propaganda media was Mariusz Błaszczak (Tab. 3). Andrzej Duda was mentioned somewhat less frequently. Both were mentioned in the context of the deployment of Patriot missiles. Mateusz Morawiecki was mentioned 26 times in the context of economic sanctions (similarly Andrzej Sadoś – mentioned 10 times). The context of Jarosław Szymczyk (17 times) was the explosion at the police headquarters and Jarosław Kaczyński was mentioned 16 times in the context of the proceedings of the ethics committee regarding the offensive comment on young women – "dające w szyję" (boozing or getting drunk).

Tab. 3. Polish politicians most mentioned on Telegram channels

Politicians	Name	Mentions	Date	Reason
Блащак	Mariusz Błaszczak	54	06.12.2022	Patriot air defence missile systems
Дуда	Andrzej Duda	42	12.12.2022	Press-conference with Steinmeier and deployment of the Patriot air defence missile systems
Моравецкий	Mateusz Morawiecki	26	15.12.2022	Sanctions (oil prices cap, 9th package)
Шимчик	Jaroslaw Szymczyk	17	15.12.2022	Explosion at Poland's national police headquarters
Ярослав Качиньский	Jarosław Kaczyński	16	13.12.2022	Kaczyński reprimanded by ethics committee for his "boozing women" comment
Лех Качиньский	Lech Kaczyński	10	15.12.2022	Polish parliament declares responsibility of Russia for the crash of Polish Air Force Flight 101 in Smolensk
Садось	Andrzej Sadoś	10	02.12.2022	Sanctions (oil price cap, 9th package)

(*Source:* original research)

The propaganda channels on Telegram quote both the Polish mainstream media (Onet, Gazeta Wyborcza, Rzeczpospolita) and the media blocked by Agencja Bezpieczeństwa Wewnętrznego (Internal Security Agency) in Poland, "Myśl Polska" and "Niezależny Dziennik Polityczny".[3]

Tab. 4. Polish mainstream, and blocked by Internal Security Agency media quoted by Telegram

Media		Mentions	Date	Topics
Polish	Interia	14	09.12.2022	Poverty in Poland, high prices on gas and electricity
	Onet	5	09.12.2022	"Michał dostał powołanie do wojska. W "kolejce" czeka nawet 200 tys. Polaków" (Michał was conscripted. Up to 200,000 Poles may be in the queue)
	Gazeta Wyborcza	37	15.12.2022	M1A2 Abrams in Poland; An article about Polish military volunteers fighting in Ukraine; Explosion at Poland's national police headquarters
	Myśl Polska	13	08.12.2022	Mateusz Piskorski: "Wszystko, co przyspiesza zakończenie konfliktu, jest dla Polski korzystne. Nawet, jeśli nie jest korzystne dla Kijowa" (Everything that accelerates the end of the conflict benefits Poland. Even if it does not benefit Kyiv) – on the missile attacks on the Ukrainian power infrastructure
	Rzeczpospolita	5	09.12.2022	MON nie chce karać Polaków walczących w Ukrainie. Ale konkretów brak. (The Ministry of Defence does not want to punish the Poles fighting in Ukraine. But there are no specifics.)
	Niezależny Dziennik Polityczny	8	05.12.2022	W całej Polsce powstaną "amerykańskie" cmentarze wojskowe ("American" military cemetaries will spread across Poland)

3 https://m.wirtualnemedia.pl/m/artykul/abw-blokada-stron-polska-rosja-ukraina-wojna-propaganda-panoptykon.

Tab. 4. Continued

Media		Mentions	Date	Topics
Other	Politico	15	15.12.2022	Poland refused to back a package deal to release €18 billion in European Union aid for Ukraine
	Stars and stripes	12	08.12.2022	"Train with the Army" project
	New York Times	9	12.12.2022	Conflicts between Germany and Poland
	Bloomberg	27	11.12.2022	Oil price cap

(*Source:* original research)

TG channels very often refer to (or pretend to refer to) some source without naming it, but using the phrase "according to Polish media".

In one such case, the TG channel published the photograph of the Arlington National Cemetery in the USA with the following caption: "The photo shows the actual military cemetery in the Polish town of Olsztyn (published by local media)" (Fig. 3).

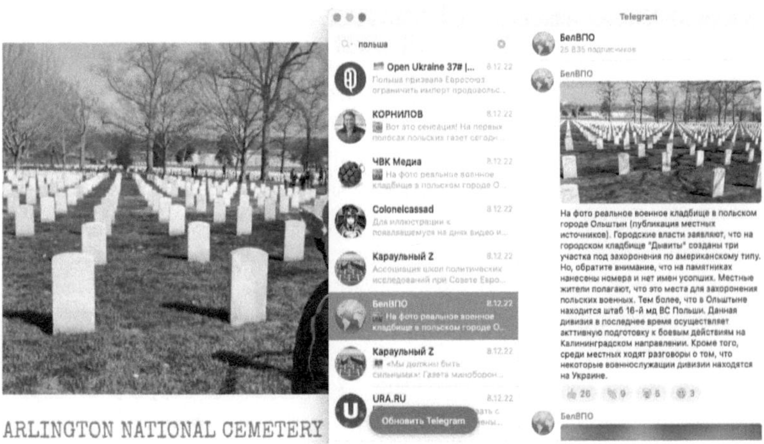

Fig. 3. Example of fake news on Telegram (*Source*: original research)

In the period from 1 to 15 December 2022, Telegram was dominated by a narrative based on the following theses: The Poles are Slavs, so they are our brothers, but they betrayed us and made an alliance with the Americans, so they are American puppets now. Poland, Hungary and Romania are preparing to annex Western Ukraine soon. We remember how we expelled the Poles from the Kremlin in 1612 and how many times we participated in the partitions of Poland. Today, the Poles are spearheading a "Russophobic alliance" with the Baltic states, against the interests of Western Europe. The Poles still argue about reparations with the Germans, so any alliance between the Poles and the Germans will not last long. Polish citizens have had enough of sponsoring Ukrainian refugees and politicians, who still admire Bandera and who want everything for free, because they are victims of Russian aggression. Prices are increasing and the Poles are heading towards mass poverty and freezing in their homes, because coal is too expensive today. At the same time, as many as 1,700 Polish mercenaries have been already killed in Ukraine, 200,000 Polish citizens are registered for the military training planned for 27 March, so it is highly probable that World War III will break out on precisely that day. However, Poland is already waging war with Russia by sending mercenaries, repairing Ukrainian military equipment and sending its own military equipment to Ukraine. Polish military aggression will be initiated soon, either towards Belarus, in the Brest oblast, or towards Russia, "from the Suwałki Gap" or "in the Kaliningrad oblast" (this makes no sense in geographic terms, but such "details" are intended to make the propaganda sound more down-to-earth than it really is).

As far as the details are concerned, short-lasting themes appear, like "the planned Polish invasion from the Suwałki Gap" or "Polish mercenaries raping Ukrainian women in Mykolovo". These are classic fake news that appear on a regular basis.

As far as hate speech ("ляхи", "пшеки" – pejorative words for Poles) is concerned, not more than 80 posts included such words (less than 0.5 %). Probably even less, because such words are not numerous and "лях" may be part of a different non-pejorative word (i.e. "бляха" ("steel sheet"), "в полях" ("in the fields")).

Life span of the news. In the posts collected on the missile attack in Vinnytsia, at least 45 reoccurring phrases were identified (due to the untypical vocabulary or specific combinations of words). The life span of such news (from the first to the last mention) was from 1 to 138 hours, 28 hours on average. However, based on the analysis of the mentions of Poland, the situation is different. Reoccurring phrases are not so apparent, they are copied more carefully, so it is more effective to analyse the themes using keywords. The themes like the adoption of packages of sanctions or discussions on the cap on oil prices may last for weeks and months. For the purposes of this study, ten themes were identified. Doubtless, the actual number of themes is much greater, but the differentiation of the themes was far more complicated than in the case of Vinnytsia. The "keywords" approach to searching was nearly futile. For example, some news reports on the incident with the Polish ambassador in Irkutsk included the event title ("Из Польши с любовью"), some included the name of the ambassador ("Краевский"), but without the event title, while search by location ("Иркутск") returned additional irrelevant results. The life span of the news was also relatively long compared to Vinnytsia: from 31 to 338 hours, 167 hours on average in the sample.

Tab. 5. Life span of selected news on Telegram

Code of event	Event	Date of event	Date start	Date finish	Lifespan (hours)
футбол	Participation of the Polish team in FIFA World Cup Qatar 2022	prior to 04/12/2022	01.12.2022 01:01:00	05.12.2022 15:06:00	110:05:00
Молдова	Missile crashes in Moldova after Russian attack on Ukraine	05.12.2022	05.12.2022 14:43:00	07.12.2022 13:56:00	47:13:00
окровавленные пакеты	Ukrainian embassies receive blood-stained parcels with animal parts	prior to 07/12/2022	02.12.2022 11:43:00	12.12.2022 11:22:00	239:39:00
кладбища	Publication of article "Niechlubny koniec. Amerykańskie kwatery dla polskich najemników" on "Niezależny Dziennik Polityczny"	23.11.2022	01.12.2022 08:50:00	13.12.2022 17:00:00	296:10:00
Марганец	RIA Novosti published a piece of news citing an anonymous source and stating that "Polish special services dressed in Ukrainian uniform, arrived in the city of Marganets, Dnepropetrovsk region – they will be looking for those who help Russia"	03.12.2022	03.12.2022 02:00:00	04.12.2022 15:44:00	37:44:00
Серадзь + 27 марта	Onet published an article "Michał dostał powołanie do wojska. W "kolejce" czeka nawet 200 tys. Polaków"	07.12.2022	07.12.2022 07:25:00	15.12.2022 20:50:00	205:25:00

(*continued on next page*)

Tab. 5. Continued

Code of event	Event	Date of event	Date start	Date finish	Lifespan (hours)
передача тел	Publication of this post https://t.me/ramzayiegokomanda/1201 about "exchanging of bodies of Polish soldiers"	07.12.2022	07.12.2022 07:40:00	08.12.2022 15:02:00	31:22:00
18 млрд	€18 billion support package to Ukraine for 2023	06/12/2022, 15/12/2022	06.12.2022 17:00:00	15.12.2022 22:45:00	221:45:00
«Из Польши с любовью»	In Irkutsk an activist presented to the Polish ambassador a picture of destroyed Soviet monuments	08.12.2022	08.12.2022 14:17:00	14.12.2022 14:09:00	143:52:00
репарации	Reparations from Germany to Poland	always	01.12.2022 08:52:00	15.12.2022 11:01:00	338:09:00
				average	**167:08:24**
Алиханов	Governor of Kaliningrad region wants to sell to Poland the construction materials needed for a fence on the border with Russia	15.12.2022	15.12.2022 05:30:00	15.12.2022 15:55:00	10:25:00

(*Source:* original research)

To conclude, the approach consisting in identifying and searching specific news on Telegram is more efficient in the case of specific and short-term themes that include multiple minor news items. This broad theme, such as the image of Poland in Russian propaganda, includes an excessive number of sub-themes (from the sanctions to the history of Polish-Russian relations since the 17th century), and each of these sub-themes includes a series of news. Searching by keywords or key phrases is ineffective.

The propaganda channels use Russian-backed media in Polish, such as Myśl Polska or Niezależny Dziennik Polityczny, to produce news "based on Polish media" and pick information from Polish media to construct a detailed image of the "terrible life in Europe". The fake news on "Polish mercenaries in Ukraine" goes from "a confessed story" to "data behind expert opinions".

There are Telegram channels specialised in information on Poland. They are not popular enough to shape opinions, but their narratives are aligned with the general narrative. The Telegram channels connected with the Belarusian government are relatively visible in the dataset.

It is quite obvious that the purpose of the Russian internal propaganda is to convince Russians that Poland is an active participant in the war against Russia, not just an ally of Ukraine. Poland is mentioned on propaganda channels on a daily basis and extensively.

Statistical Analysis

As a result of research, 1,665 posts were collected, and the corpus wordcount was 12,019. The most frequently used word was "Poland", with the share of 2.77 % in the corpus. The next were "Ukraine" (2.30 %) and "Russia" (1.25 %). In the 25 most common words, the sixth was "USA" (0.36 %), the seventh was "Germany" (0.33 %), and the eighth was "Europe" (0.33 %). These 25 words form as much as 12.80 % of the analysed texts. However, the notion that 20 % of the lexis creates ca. 80 % of the content remains consistent with the proportions defined more than 100 years ago by Vilfredo Pareto (Fig. 4).

It is not surprising that the most frequently occurring lexis includes terms related to aggression and states directly and indirectly involved in it (Plot 1).

Tab. 6. Frequency of words in the collected corpus

No.	Words	% in total	% total
1	Польск	2.77	2.77
2	украин	2.30	5.08
3	росс	1.25	6.33
4	стран	0.82	7.15
5	воен	0.49	7.64
6	сша	0.36	8.00
7	герман	0.33	8.34
8	европ	0.33	8.67
9	поляк	0.33	9.00
10	цен	0.31	9.30
11	нат	0.29	9.59

(continued on next page)

Tab. 6. Continued

No.	Words	% in total	% total
12	заяв	0.27	9.86
13	территор	0.27	10.12
14	беженц	0.25	10.37
15	войн	0.25	10.62
16	границ	0.24	10.86
17	удар	0.24	11.10
18	западн	0.23	11.33
19	запад	0.23	11.56
20	наш	0.22	11.77
21	русск	0.22	11.99
22	нефт	0.21	12.20
23	возможн	0.21	12.41
24	оборон	0.20	12.61
25	европейск	0.19	12.80

(*Source:* original research)

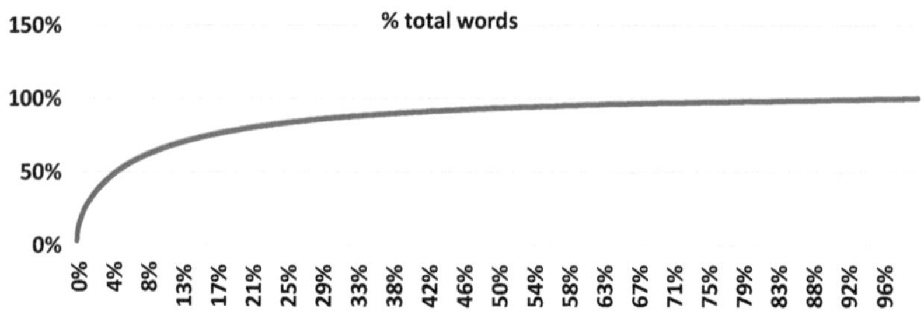

Fig. 4. Lexis vs content (*Source: original research*)

Plot 1. Cloud of lexis in the studied corpus (*Source:* original research)

In the content collected using statistics software, 18 dominant themes were singled out (Tab. 7). The most frequent theme, which occurred as much as 208 times, was connected with Russia, Ukraine, Poland and war.

Tab. 7. Singled out themes and their frequency

No.	Topic	Count	Name
1	0	208	0_россия_украина_польша_война
2	1	180	1_военный_польша_танк_польский
3	2	135	2_беженец_украинец_польша_помощь
4	3	124	3_германия_patriot_система_польша
5	4	94	4_нефть_цена_потолок_российский
6	5	89	5_посол_польский_польша_выставка
7	6	73	6_россия_украина_запад_польша
8	7	61	7_полиция_подарок_комендант_взорваться
9	8	58	8_санкция_пакет_ес_страна
10	9	51	9_терроризм_сейм_резолюция_спонсор
11	10	43	10_электричество_генератор_квт_польша
12	11	38	11_посольство_пакет_животное_глаз
13	12	35	12_газ_цена_ес_снизить
14	13	26	13_зеленский_президент_мировой_издание
15	14	22	14_декабрь_бразилия_аргентина_португалия
16	15	20	15_18_помощь_евро_будапешт
17	16	20	16_ракета_молдавия_упасть_300
18	17	18	17_сборная_франция_матч_обыграть

(*Source:* original research)

The dominant theme, which occurred as much as 208 times, was Russia, Ukraine, Poland and war. In the narrative of the studied entries, Poland was positioned similarly to Ukraine and even if it was not directly involved in military operations, it was regarded similarly to Ukraine.

The strong involvement of Poland was confirmed by the themes related to war, Poland and tanks, which occurred 180 times in the studied texts (referring both to the deliveries of the Polish Leopard tanks to Ukraine and the pressure exerted by the Polish government on other NATO member states to deliver such arms). Another frequent theme (135 times) is Poland, Ukrainian refugees and aid to them. A significant share in the narrative was the theme of armament – the theme of German Patriot missiles and their deployment in Poland occurred 124 times.

The list of the six most frequent themes is closed by Russian oil prices, 94 times, and the Polish exhibition in Irkutsk, 89 times. The most frequent lexis clouds are presented in Fig. 5.

Fig. 5. Lexis clouds of the six most frequent themes (*Source:* original research)

The above themes occurred at various rates (Tab. 6). The first three of the six most frequent themes were addressed evenly the entire studied period and clearly dominated the following three (Figs 6 and 7).

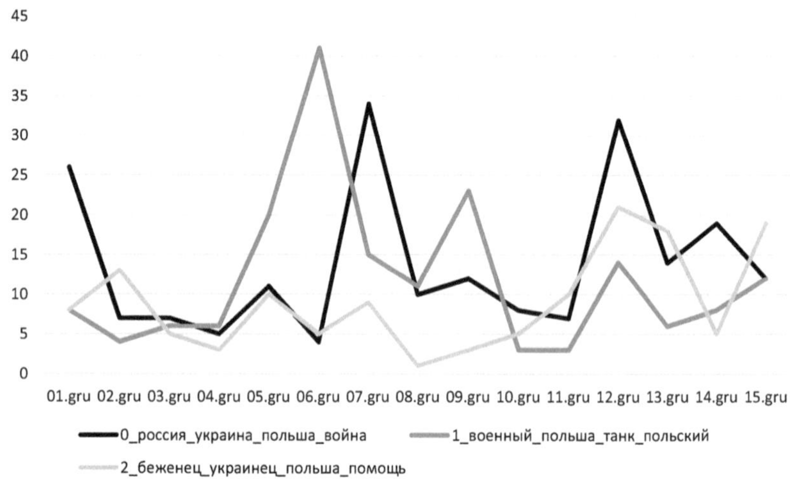

Fig. 6. Occurrence of themes in the period of 01–15. 12 (*Source:* original research)

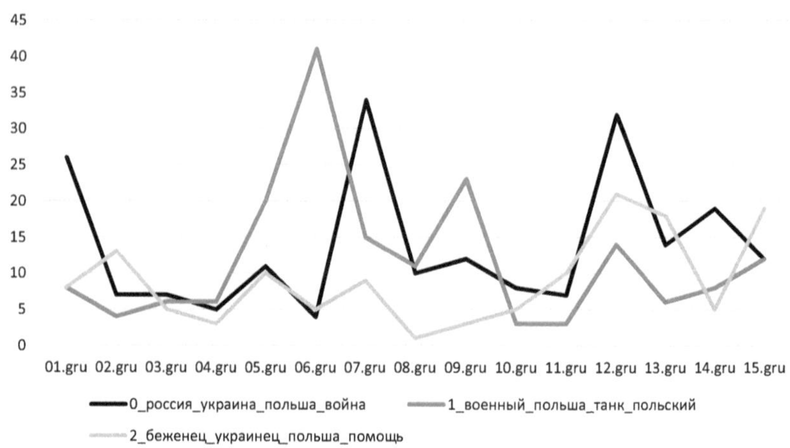

Fig. 7. Occurrence of themes in the period of 01–15. 12 (*Source:* original research)

Tab. 8. Occurrence of themes in the period of 01–15.12.2022

Themes	01. Dec	02. Dec	03. Dec	04. Dec	05. Dec	06. Dec	07. Dec	08. Dec	09. Dec	10. Dec	11. Dec	12. Dec	13. Dec	14. Dec	15. Dec
0_россия_украина_польша _ война	26	7	7	5	11	4	34	10	12	8	7	32	14	19	12
1_военный_польша_тан к_ польский	8	4	6	6	20	41	15	11	23	3	3	14	6	8	12
2_беженец_украинец _ польша_помощь	8	13	5	3	10	5	9	1	3	5	10	21	18	5	19
3_германия_patriot _система_ польша	10	4	0	3	1	34	17	0	0	10	3	6	30	2	4
4_нефть_цена_потолок _ российский	28	29	19	4	8	1	0	0	0	0	1	0	0	0	4
5_посол_польский _польша_ выставка	4	2	5	1	5	4	6	12	9	9	8	9	7	3	5

(*Source:* original research)

Conclusion

In the period studied, Poland occurred mostly in the context of the Ukrainian-Russian war. War, tanks and refugees shaped the content of these narratives. It should be also noted that the statistically dominant themes coincided with the identified themes of fake news. The authors of the fake news reports created them based on true information. It should be also noted that the large number of news reports unfavourable to Poland resulted from the intense exploitation of accurate information. And, for example, in case of the firing of the grenade launcher in the office of the police chief was shown intensively, which made possible the hypothesis that Polish public information policy is weak. Unfortunately, such content creates the image of Poland as a weak state. In addition, the ongoing narratives position Poland in the middle of the conflict or even at as a party to it.

Katarzyna Materska

Contributions of Polish Projects to Countering Disinformation (2019–2022)

The Complicated Landscape of Information Disorder: Instead of an Introduction

Today's world is dominated by digital information exchange. It is estimated that ca. 60 % of the human population (more than 5 billion people) are Internet users [15]. As the use of cyberspace develops, we may see increasing disruptions in digital ecosystems, which entail disturbing hazards for their users (either occurring as a natural consequence of the scale and rate of the ongoing changes, or triggered artificially for a specific purpose). This is undoubtedly caused by both people and technologies.

The notion of information disorder (information distortions, pathologies, deviations) is usually associated with the very influential report by Claire Wardle and Hossein Derakshan for the Council of Europe in 2017 [5, 6], and the subsequent publications by these authors [4, 7], in which they proposed a typology defined around three main types of information disorder: disinformation, misinformation and malinformation, which often makes use of true information. These types are characterised by different degrees of wilfulness and intention to cause harm. However, it should be noted that according to some researchers, this feature is a limitation, because the motives are often diverse, contrary or unclear (for more see Ref. [8], p. 9).

In view of the above, it is easy to note that the notion of information disorder, taken as defined, has appeared in research (e.g. by the environments dealing with media and information competences) relatively recently, no earlier than 2017.

Information disorders may also be understood more broadly, as a large-scale disruption of the public sphere, caused by gossip, disinformation, fake news, hate speech, dangerous and unfounded conspiracy theories, harmful misunderstandings and planned fraud campaigns etc. The COVID-19 pandemic and the associated "infodemic" of harmful myths and theories, and false treatments that are dangerous to health and life, as well as interference in the election campaigns in many countries have shown the high degree to which these phenomena can destabilise the normal functioning of units, groups and societies, even on a global scale; the events of the war in Ukraine confirm that disinformation may be used as an effective weapon. Information disorders are often regarded as a conglomerate of quickly changing media ecology and the increasingly fractal, populist and decentralised political environment (see Ref. [3], p. 1). The "fuel" for spreading lies, manipulations and propaganda are to a large extent social media.

Manipulating, hiding, withholding or silencing the truth and lying have been recurring phenomena in the entire history of humankind. However, the tools and technologies that it has acquired over the last three decades have contributed to the unprecedented rate and scale of changes in the information and communication space, which have somehow got out of control (the original intent having been their use for creating a friendly information environment) and become tools of the different actors of the broadly understood disinformation for their business, political or other purposes, sometimes evil.

In general, information disorders related to current affairs are often referred to as fake news. They very often look like true information, however following the analysis and verification of sources, they prove to be information disorders. Considering that the term "fake news" (false reports) is not always synonymous to "disinformation", in this article "disinformation" will be used as a general term designating the contents that are false, distorted, inaccurate etc., and thus having a potentially negative impact, regardless of whether the disinformation is intentional (fake news) or unintentional (e.g. the so-called urban legends).

At this point, it is worth noting two perspectives in the approach to disinformation. In qualitative terms, disinformation is understood as information with the capacity of creating a manipulated image of reality for a specific purpose. Disinformation is treated as information disorder in the so-called static approach. In the dynamic approach, disinformation is treated on a process and intentional basis, namely as a "process of manipulation, wilful misrepresentation, causing doubt as to the factual state of affairs in the audience by means of different types of information disorders" [14]. The purpose of such information operation is to interfere with the process of reaching situational awareness, creating a specific image of reality in the audience, and by consequence a decision (or the lack of it), as planned by the entity planning the information operation. "Disinformation in itself is interference with the decision-making process of the audience (individual or group), also by masking certain events and information on such events or generating information noise to confuse the audience. Thus, the disinformation process uses information that is not always false, and it includes an entire range of information disorders" [14]. Rafał Kasprzyk [14] points to the great similarity of the disinformation process to marketing activities, including advertising activities.

The increasing availability of technology contributes to the continuous evolution and refinement of ideas of the participants of information environments, their tools, strategies and tactics in the field of disinformation. It may even be tempting to speak of a high level of "innovation" in the field of disinformation, not always in a positive sense. For this reason, different indications related to fake news and disinformation are concealed with increasing cunning and more difficult to detect, while great numbers of common mortals are not even aware that the decisions they make are not fact-based decisions.

Speaking of the disinformed and the manipulated, different scales of entities must be considered – from individuals to specific social groups, to even entire countries and international communities. In the globalised world, which is increasingly often studied as a set of different ecosystems, it must not go unnoticed that changes in some components (e.g. in technologies) must have an impact on other components (e.g. behaviours, emotions and values of people).

The struggle against disinformation involves different resources. Speaking of technology, it must be noted that the situation is quite paradoxical. On one hand, so-called artificial intelligence (AI) systems aggravate the problems of disinformation (by reinforcing the phenomenon of disinformation) that we are facing at present. Firstly, they increase the capabilities for creating increasingly realistic fake contents, manipulate texts and images, audio or video content, but secondly, or maybe above all, they facilitate the dissemination of disinformation among its (target) audience on a large scale (due to the algorithms developed and implemented by the Internet platforms, of which the purpose was to increase the engagement of their users). This image is supplemented by the entire farms of bots and cyborgs. The world of disinformation is becoming even less transparent due to the significant progress in the utilities of natural language processing (NLP), which are perfect in dealing with short texts at the present stage. In addition, the open character of many mechanisms of disinformation detection that have already been developed results in a situation, in which the actors of disinformation learn and improve their tools on a continual basis. Of course, these are only examples of the negative side of the impact of the technology [1, 2].

In response, different artificial intelligence systems are developed to detect and moderate disinformation on the Internet, because the consequences of its dissemination are increasingly alarming. Here moderation means, for example, filtering, ranking, blocking or removing content or disabling and suspending accounts, which the technology providers may use to mitigate disinformation on the Internet.

At present, there are many activities undertaken worldwide to fight disinformation (including fake news). The reasons for the intensification and acceleration of these activities were undoubtedly the COVID infodemic and the war in Ukraine. To counteract these negative phenomena, legal regulations are introduced and IT and media education is implemented on an increasingly wide scale to support the users in the network.

Polish scientists too have been entering thus outlined landscape for a few years, with the purpose of finding the best remedial solutions.

Presentation of Selected Projects

Eleven Polish research projects involved in the fight against disinformation were selected for study. Namely[1]:

- **HOMADOS**. Hampering misinformation by assessing credibility of online sources – project website: https://homados.ipipan.waw.pl/,
- **#Cyber_odporność**. How can Internet social media build resistance to disinformation? – project website: https://webimmunization.cm-uj.krakow.pl/,
- **IKONA.** Identification, collection and evaluation of hostile disinformation operations in cyberspace, conducted based on telecomputing utilities – project website: https://www.inforafinacja.pl/projekt-ikona/,
- **SWAROG.** System for detecting disinformation using artificial intelligence methods – project website: https://www.kssk.pwr.edu.pl/projects/swarog,
- **OpenFact.** Utilities for checking the credibility of information sources and detecting false information using artificial intelligence methods – project website: https://kie.ue.poznan.pl/pl/openfact/,
- **AntyFakeNews**. System for protecting users from false information distributed on the Internet – project website: https://itd24.pl/o-projekcie/,
- **From urban legend to fake news.** The global detector of modern falsehoods – project website: https://american-systems.pl/ncbr-infostrateg/,
- **Detection of manipulated audio-video content** to prevent the dissemination of deepfake news [no project website],
- **Intelligent System of Discursive Fake News Detection** [no project website],
- **INFOTESTER.** Development and verification of original vertical artificial intelligence methods for automatic and precise detection of disinformation – project website: https://infotester.pl/,
- **NONSENSE AND PROPAGANDA Online**. Internet communities and bots propagating misinformation (no project website), general project characteristics

1 Further in this article, the abbreviated name of the project (marked in bold) will be used. Not all projects have a project website (as of the end of December 2022).

available at: https://www.kozminski.edu.pl/pl/nonsens-i-propaganda-online-spolecznosci-i-boty-propagujace-dezinformacje-w-internecie.

The criteria for selecting the projects for the study were: nationwide sources of funding, current implementation period (2019–2022) and involvement of academic research groups.

Here it is worth noting that in addition to the projects listed, Polish research teams are also involved in different international projects that are not listed (e.g. DISSIMILAR – project website: http://dissimilar.ii.pw.edu.pl/). Many projects are also pursued with the tasks defined above commissioned by individual ministries, for example, the Ministry of Health (MEDFAKE): Building Trust in Vaccination using the technically advanced communication tools and social impact methods. The Consortium includes The Medical Postgraduate Training Centre (Leader), the SWPS University of Social Sciences and Humanities, the Medical University of Warsaw, the Kozminski University. In addition, individual universities appoint expert groups, for example, the Research Team for Propaganda and Disinformation appointed in 2020 at the Institute of Political Sciences and Administration of UMCS in Lublin. The team issues "Monitoring Propagandy i Dezinformacji" (Monitoring of Propaganda and Disinformation) (see e.g. https://phavi.umcs.pl/at/attachments/2021/0224/125911-biuletyn-politologiano2-9cz-1.pdf). An example of a Polish social initiative could be #FakeHunter – a social project of verification of contents published on the Internet launched by the Polish Press Agency and GovTech Polska, aimed at debunking false information on the SARS-CoV-2 virus.

Returning to the characteristics of the projects selected for this study, they are projects currently being pursued (except for one, the HOMADOS project ended in the year of writing). The project implementation periods are as follows (ordered by planned ending):

ended:
- **HOMADOS** (January 2019–March 2022)

in progress:
- **#Cyber_odporność** (25 September 2020–24 March 2023)
- **IKONA** (1 September 2021–30 June 2024)
- **SWAROG** (1 December 2021–3 March 2025)
- **OpenFact** (2 December 2021–31 March 2025)
- **AntyFakeNews** (2 December 2021–31 March 2025)
- **From urban legend to fake news** (December 2021–March 2025)
- **Detection of manipulated audio-video content** (2 December 2021–1 June 2025)
- **Intelligent System of Discursive Fake News Detection** (2 December 2021–1 June 2025)

- **INFOTESTER** (2 December 2021–1 June 2025)
- **NONSENSE AND PROPAGANDA** (30 November 2021–29 November 2026).

The fact that the tasks under most projects were undertaken in 2022 contributes to the fact that not many materials on the progress of works have been released to the public, as they are still at the initial stage of implementation. The basis for the analysis presented here is the general project information declared in the applications to the grant funders (e.g. the National Centre for Research and Development, the National Science Centre) and the information shared on the individual project websites.

Tab. 1 shows that most projects were funded by the National Centre for Research and Development (NCBiR) under the INFOSTRATEG programme, the strategic programme of scientific research and development projects. In the first year when a call for proposals was issued, addressed to entrepreneurs and scientists, two specific research topics were presented, including: *Verification of sources of information and detection of fake news*. The objective specified for the projects completing this subject was to build utilities to detect false information and to enable marking this information in search engines, social media and other popular communication and information dissemination systems. The comparison of the number of properly found false pieces of information was to be the basis for the final evaluation. The grant funder hopes that the task will contribute to the reduction of the negative impact of fake news and restrictions of their dissemination, and that the creation of such system could restrict the attempts to manipulate the public opinion. Proposals were accepted from 4 January 2021 to 3 March 2021.

The National Centre for Research and Development also acts as the grant funder of the CyberSecIdent programme – the research and development programme "Cybersafety and e-Identity", pursued in collaboration with the Ministry of Digitalisation. The funding was granted to the IKONA project (subject area: Technologies and solutions for the detection, presentation and protection from hazards in cyberspace and their consequences at the state level).

Tab. 1. Project grant funders and funding amounts

Grant funder	**Project name**	**Subsidised funding amount**	**Total project budget**
NAWA, Google Cloud and PCSS calculation grants	HOMADOS	USD 5,000, PLN 18,800 PLN, 50,000 h of calculations in PCSS	

Tab. 1. Continued

Grant funder	**Project name**	**Subsidised funding amount**	**Total project budget**
NCBiR, INFOSTRATEG I contest	SWAROG	PLN 7,935,166.25	PLN 8,657,006.25
	OpenFact	PLN 3,400,500	PLN 3,400,500
	AntyFakeNews	PLN 3,265,794.85	PLN 4,177,959.56
	From urban legend to fake news	PLN 6,641,828.82	PLN 7,907,896.47
	Detection of manipulated audio-video content	PLN 3,965,841.58	PLN 3,965,841.58
	Intelligent System of Discursive Fake News Detection	PLN 4,576,587.12	PLN 6,366,399.85)
	INFOTESTER	PLN 7,898,438.23	PLN 8,782,106.67
NCBiR, "CyberSecIdent – Cybersecurity and e-Identity"	IKONA	PLN 6,605,669	PLN 7,926,959
EOG Financial Mechanism for 2014–2021 – under the IdeaLab contest, NCN	**#Cyber_odporność**	PLN 5,667,500	
NCN, 2021 Maestro grant	NONSENSE AND PROPAGANDA	PLN 3,807,634	

(*Source:* own study)

The formal characteristics of the presented projects is complemented by the "scientific background" of research teams.

- **HOMADOS** – Team of Linguistic Engineering at the Institute of Computer Science PAS;
- **#Cyber_odporność** – coordinator: Jagiellonian University, Collegium Medicum; partners: University of Oslo, Faculty of Psychology, Poznań University of Technology, Faculty of Computer Science and Telecommunication; Maj Institute of Pharmacology PAS (Kraków);
- **IKONA** – Centrum Rafinacji Informacji sp. z o.o. (spin-off UW), University of Warsaw (UW);
- **SWAROG** – coordinator: Faculty of Electronics, Wrocław University of Science and Technology (Department of Systems and Computer Networks); partners: MATIC S.A., Jan and Jędrzej Śniadecki Bydgoszcz University of Technology and Agriculture (Faculty of Telecommunications, Computer Science and Electrical Engineering);

- **OpenFact** – Poznań University of Economics and Business (Department of Information Economics);
- **AntyFakeNews** – partners: WSB University; coordinator: ITD24 Sp. z o.o.;
- **From urban legend to fake news** – partners: Adam Mickiewicz University in Poznań, Centre for Artificial Intelligence, coordinator: American Systems Sp. z o.o;
- **Detection of manipulated audio-video content** – Cracow University of Technology;
- **INFOTESTER** – Medical University of Białystok, Polish-Japanese Academy of Information Technology;
- **NONSENSE AND PROPAGANDA** – Project manager: prof. dr hab. Dariusz Jemielniak (Kozminski University – ALK).

Only one project considered in the article, the Intelligent System of Discursive Fake News Detection, is pursued without a scientific partner, i.e. by the technological start-up Swarmcheck Sp. z o.o. (https://www.swarmcheck.ai/glowna).

Project Objectives

All project teams focus their efforts on solving the problems entailed by the contemporary information disorders. However, the main objectives specified for the projects focus on different areas, components and tasks, such as:

- Automatic detection of untrustworthy and fabricated text content (*fake news*) based on their linguistic features; automatic suggestion of credible sources relevant to the local context; implementation of the prototype of an easy-to-use tool that will help non-expert users (HOMADOS);
- Restriction of dissemination of disinformation – study of individual susceptibility to disinformation in the network and behaviours accompanying its users on Twitter [disinformation treated as type of "disease"]; preparation of ethical principles, which could in future be an important contribution to the development of utilities necessary for moderating the public debate in the modern digital Internet society (#Cyber_odporność);
- Development of the information system to identify hostile disinformation operations in cyberspace (fake news) and their sources. IKONA will be an information system operating in real time, using big resources of unstructured data (BigData) (IKONA);
- Detection and classification of disinformation using artificial intelligence methods; web apps and plug-ins for Internet browsers supporting the user (SWAROG);

- Development of the tools to check the credibility of information sources on the Internet and detect false information, so-called fake news, using artificial intelligence methods (OpenFact), for the Polish language;
- Warning (protecting) users against potentially dangerous fake news content using the telecomputing utility. The utility will enable the user to evaluate both the content already available and to be published (sent); helping the user in the independent analysis and acquisition of competences to verify information in the network (AntyFakeNews);
- Analysis of false content on the Internet (fake news) as a basis for further empirical research on their significance. The solution is designed for public bodies (Ministry of Internal Affairs and Administration, Ministry of Defence, Police, Internal Security Agency or Military Counterintelligence Service – in terms of public safety), and in general and economic applications for publishers, the Warsaw Stock Exchange, Financial Supervision Authority (to control the manipulation of values of specific companies or national macro-economic situation), fact-checking institutions and analytical companies. [Here the solutions were created for an institutional user] (From urban legend …);
- construction of a system for detecting false information (accessible in open access mode), with justification of such evaluation in the form of argumentation graph (the Intelligent System of Discursive Fake News Detection);
- development of methods of detection of manipulated video content with deepfake characteristics using machine learning methods related to or identical with the methods currently used to produce deepfake materials; automated verification of video and audio content for credibility – the application will determine the manipulation probability on a scale [0,1] (Detection of manipulated audio-video content…);
- development and verification under laboratory conditions of quasi-real and real prototypes of three components: (1) automated evaluation of the credibility of website contents, (2) crowdsourcing evaluation of the credibility of content, based on the automated recommendation and reputation process, protecting the evaluation from "crowd" manipulations, (3) evaluation of credibility of images (INFOTESTER);
- Study of major social changes; "analysis of network, structure and organisational culture of Internet disinformation communities (in particular: their language, social structure, leadership, demographics, reputation systems, internal logic, beliefs, narratives and myths)"; Development of research tools for the Polish scientific community; Creation of a tool for detecting networks of manipulators (hired professionals and bots) (NONSENSE AND PROPAGANDA).

From the cumulative summary of project objectives, the following objectives appear: detection (development of the identification system) of manipulated/false information, video contents (deepfake); analysis (of contents, networks, communities, structures); classification of contents; evaluation of credibility (automatic, individual, crowdsourced, in the form of argumentation graph); verification (of sources, contents, images, videos); automatic suggestions, recommendations; restriction of spreading of disinformation, fake news; warning of the users of networks, social platforms; examination of the susceptibility of individuals to disinformation; development of ethical principles connected with cyber-resistance (e.g. how much can the autonomy of individuals be influenced to restrict the dissemination of fake news and to moderate the public debate in a modern digital community).

Arguably, it may be said that it is both about the digital protection of network actors (assisted by innovative tools prepared using state-of-the-art artificial intelligence methods, deep learning). The indirect proposed objective of the proposed solutions is to aid the education of users (on any scale) in the evaluation and selection of the most effective verification criteria for sources of information, text messages, images and video and audio materials.

The Object of Study in Projects

Different project teams focus on different components of the disinformation ecosystem. These are the content and form (format) of messages, sources of disinformation and participants of cyberspace.

Content[2]

In relation to the content of messages, the tasks to complete under the projects are defined as follows: detection of untrustworthy content (on the Internet and in social media), not only text content but also illustrations, which provoke more emotional reactions and mislead the reader; evaluation of credibility of statements published in the public space – development and use of advanced NLP algorithms and machine learning of automated detection and classification of fake news (in English and Polish); analysis of content dissemination; promotion

2 At the outset, it is necessary to explain two Polish notions, *treść* and *zawartość* (English: content), which are not synonymous. *Content* is not only the information (*treść*), but also other components that are sometimes important for using it or simply helpful. It may be the layout, the sequence or the interconnections (or lack thereof). This distinction is particularly visible in relation to the message content in the form

of credible content; evaluation of the truthfulness degree or the degree of manipulation, falsehood and emotional markedness of the message.

For greater legibility, an attempt to conclude the most important research on disinformation messages can be made in several categories (facets), as shown in Tab. 2.

Tab. 2. Facet taxonomy of disinformation messages as subject of study

CONTENT (INFORMATION)	
FORM	• text (language) – e.g. stylometric features • illustrations, graphics • video (deep fake)
ATTRIBUTES	• incredibility/ falsehood/ manipulation • truthfulness/ credibility • emotionality • agency (force of impact, force of effect)
LOCATION	• Internet • social media
PROCESSES	• sharing (wilful) • dissemination (also unintentional) • social control processes • promoting (e.g. of credible contents)

(*Source:* own study)

Those executing the project do not for the most part restrict the scope of content to any closed, selected theme. The analysis of content covered the messages in English and in most projects also in Polish (e.g. Open Fact, IKONA).

Sources of (Dis)information

The consideration of the sources of (dis)information was specifically indicated in three projects: HOMADOS – as suggested by the project title itself (Hampering Misinformation By Assessing Of Online Sources); verification of credibility of the sources of information is highlighted under the OpenFact project and identification of sources of disinformation is highlighted under the IKONA project. The examples may indicate two different approaches to sources of information. In the first case, the source of disinformation is understood statically in the form

of image or video (in which the artistic form matters just as much as the intellectual content). In the projects discussed in the article, the term *treść* (content) is used.

of a specific distorted message on reality (result of action), while in the second case the focus is on the source of disinformation operations, that is, the source of an entire series of actions related to the intentional impact of specific entities (dynamic approach). What matters here is not only the content (result) but also the intention (intention of the creator), that is, the disinformation context. In the implementation of further project phases of the ongoing research, the approach to sources of disinformation will be verified. Both the identification and the evaluation of credibility and verification of sources are essential for fighting disinformation.

Participants in the Disinformation Ecosystem

In the disinformation ecosystem, diverse participants (actors) play an essential role with their roles and tasks. The potential "victims" of disinformation may be entities of varied (any) scale – from persons, companies, up to and including countries. This broadly understood recipients are addressed by the propositions of the IKONA project. Institutional users are specifically indicated in the following projects: From urban legend, IKONA. Disinformation communities are the subject of special interest in the NONSENSE AND PROPAGANDA project. However, the predominant thinking in the projects is of individual users as struggling with difficulties caused by false and manipulated information on a daily basis. The researchers of the #Cyber_odporność project are specifically interested in the mechanisms of individual susceptibility of the users to disinformation in the network, while they refer to resistance to fake news as cyber-resistance. Considering the behaviours and emotions of users in the network, they try to create a psychological profile of the persons most susceptible to believe in false information, in order to learn how to efficiently suppress fake news epidemics as they start. This is connected with the identification of the "social DNA of fake news", in order to specifically examine the functioning mechanisms of disinformation in different groups of media users, describe and classify the models considering the levels of susceptibility to specific types of false information, awareness of their meaning, methods of referring to fake news, mechanisms of protection against them, as well as specifics of their dissemination in the space of network media (AntyFakenews).

However, the users of the disinformation environments are not only "victims", but also the authors and broadcasters of content (including distorted messages), organisers of disinformation campaigns, and creators of conspiracy theories, who have different goals and pursue various strategies. This group of actors of the disinformation (cyber)space, both humans (including hired professionals) and bots, is examined under the NONSENSE AND PROPAGANDA project. The project executor is interested in a deepened analysis of the organisational network, structure and culture of selected Internet disinformation communities.

The detection of untrustworthy users (bots) is the purpose of the HOMADOS project team.

Methods Adopted in Pursuit of the Objectives

Specific methods in pursuit of the objectives are based above all on the selection of the main object of study. As outlined above, different components of the world of disinformation were selected in the pursuit of the objectives. In addition, the methods in use are shaped by the interdisciplinary nature of project teams, including members from different specialities. For example, the "From urban legend to fake news. The global detector of modern falsehood" project involves mathematicians, computer specialists, linguists, media specialists, lawyers, philosophers, folklore specialists and IT specialists. Therefore, it is no surprise that the team built in this manner applies both methods in the humanities and neural networks, Natural Language Inference technologies, REALM-type models etc. The project website indicates the combination of narratological, comparative and sociological methods with computer NLP methods and big data analysis, as well as machine learning methods.

The #Cyberodporność project team is interdisciplinary as well. It employs ethicists, psychologists, neurobiologists, epidemiologists and computer scientists, who work to investigate the mechanisms of the "disease", which is disinformation. In order to build the scale of susceptibility to disinformation, for example, psychometric tests were used.

The OpenFact team states their application of data science methods, including among other things the technologies of (deep) machine learning, artificial intelligence and natural language processing. Similarly, the IKONA project uses large sets of unstructured data (BigData), artificial intelligence algorithms and advanced algorithms based on operation studies, machine learning and statistics. In turn NLP, social media analysis and machine learning (mostly in data classification), advanced supervised learning models, as well as classifier teams are found among the HOMADOS and SWAROG project methods.

The NONSENSE AND PROPAGANDA project methods were also adjusted to the object of study. These were described as Thick Big Data methods, which combine the Big Data approach (TribeFinding, SNA, Machine Learning, Sentiment Analysis) and qualitative approaches (digital ethnography, narrative analysis).

Obviously, the methods applied in the projects are also connected with planned utilities. The examples of utilities that achieve the objectives include BotHunter, for the detection of bots on Twitter (HOMADOS); the development of a new international bot detector (NONSENSE AND PROPAGANDA); Credibilator (browser extension) for automatic evaluation of credibility of websites currently viewed (HOMADOS).

Difficulties in the Completion of the Studies

Certainly, detailed information on the difficulties is found in project summary reports. It seems, however, that two essential matters arise from the current information. Firstly, already at the initial stage of research, many teams encountered difficulties related to the problematic nature of the very notion of disinformation, fake news etc., which was highlighted in the introduction to this article. The findings on the scope of notions and terms in use are essential, as they imply an entire series of subsequent assumptions, for example, determining whether something is disinformation or not only in true and false categories or considering (or not) the intention of the persons forwarding the content and other contextual circumstances. Secondly, the problems defined in the "social or cultural context, including the detection of disinformation, lead to the development of models, of which the maximum efficiency is achieved at the moment of collection of the training data set. Any such system is already outdated the moment it is published. In use, it is subject to further degeneration, in parallel and proportionally to changes in class properties for dynamic recognition, referred to as the drift of notions" (SWAROG).

Contributions of Polish Projects

The struggle against the global problem of disinformation and the search for the best protection against information threats in cyberspace are on the list of priorities of modern societies. Therefore, the projects undertaken by Polish teams are up-to-date and socially necessary. Social media gave users the opportunities to express themselves and articulate their opinions and views, as well as expectations of approval or even accolades. Such attitudes create an environment dominated by emotions, in which substantive discourse is often difficult to find. In addition, different interest groups are increasingly active, using disinformation and disinforming for their own benefit. Dariusz Jemielniak, in the NONSENSE AND PROPAGANDA project, noticed three major social changes:

1. increasing distrust in institutional knowledge and anti-intellectual and anti-establishment social sentiments;

2. increasing tendencies of Internet communities to self-organise around specific ideas, as well as new forms of online cooperation (partner production, open cooperation), combined with democratisation of the creation and propagation of knowledge, which result in disinformation;
3. an increasing role of automatic propaganda and disinformation utilities (networks of bots, paid undercover influencers, self-propelling networks of professional ideologues).

An advantage of the Polish projects is the diversity of approaches, conceptions and perspectives. Each project team focuses on different components of the disinformation ecosystem. Covering them with cumulative attention may contribute to finding the interdependency of the individual components of the complex world of disinformation. Better knowledge of the nature of disinformation, fake news and other information disorders, and identification of their attributes and the mechanisms of their creation, dissemination and impact on different fields of human life (on an individual, group and global level) may lead to the creation of technological utilities and social mechanisms to help these actors of the world of information that feel increasingly lost in the ocean of (dis)information. This is also related to the identification and prevention of the practices used by disinformation actors, by "Internet communities and bots propagating misinformation" (NONSENSE AND PROPAGANDA). In the projects discussed here, the diverse methodology should contribute to the expected results – quantitative and qualitative approaches (e.g. digital ethnography, narrative analysis, computer linguistics, artificial intelligence and Big Data). Moving beyond text contents and local context (e.g. language) should be considered important.

It is important to note that disinformation causes other hazards for entities subject to it on different scales, and thus other solutions are necessary. For example, disinformation to an individual may be a disease, an infringement on their subjectivity or human dignity (e.g. in the #Cyberodporność project). The support developed by ethicists, psychologists, neurobiologists, epidemiologists and computer scientists must be recognised as being of incalculable value. At the level of a spontaneous social group, disinformation may be viewed as grounds for questioning the existing system of knowledge, as the component bonding the members of specific Internet disinformation communities, influencing their structure and organisational culture (e.g. leadership, language, beliefs, narratives and myths). This requires combining computer science, management, sociology and ethnography (NONSENSE AND PROPAGANDA). At the level of states and the international community, disinformation creates threats to democratic processes and world peace; it is used as means of achieving gain and economic, political and military advantage. The solutions proposed at this level are intended for public entities, for example, in the field of public safety, to control the manipulation of value of specific companies or the national macro-economic situation (from urban legend…).

The interdisciplinary nature of the above teams contributes to the application and development of many broad perspectives – linguistic, cultural, psychological, sociological, philosophical, epidemiological, computer science and others. One should hope that such a holistic and simultaneously multi-dimensional perspective will prove the strength of the efforts undertaken and solutions proposed, creating a basis for further deepening of future research work.

Conclusion

The disinformation ecosystem is a highly dynamic field, in which the pace and scale of changes is applicable to all of its components: utilities and technologies, contents, applied methods and strategies, as well as profiles of the participants (individual and collective, people, bots and cyborgs). In an environment, in which false news factories arise, facts compete with "social media facts". Facts are skilfully taken out of context, often confirmed scientifically, are used as weapons in the service of manipulation and falsehood, algorithmic recommendation systems are overused etc. Systematic (incessant) and careful monitoring, analysis and creation of diverse and courageous scenarios are required. This is the only way to minimise the danger that science will always remain one step behind reality and that the asymmetry of information and knowledge in the struggle against disinformation will deepen. At present, we do not have a "silver bullet" and it should not be believed that an individual team can develop and implement an extremely effective idea. Therefore, it is a good thing that information disorders are addressed by hosts of researchers and entrepreneurs worldwide. The people fighting disinformation are increasingly well-armed with powerful digital tools based on artificial intelligence. The question is what will the user do upon receipt of the results of evaluation of credibility of the sources and message in social media and other cyberspace services? Because the user often does not want to find the truth, but to confirm their own beliefs.

References

[1] Beyer, Jan Nicola, Lena-Maria Böswald. "On the radar: Mapping the Tools, Tactics and narratives of Tomorrow's Disinformation Environment". Democracy Reporting International, June 2022. https://democracyreporting.s3.eu-central-1.amazonaws.com/images/62c8333ec3aea.pdf.

[2] Bontridder, Noémi, Yves Poullet. "The role of artificial intelligence in disinformation". Data & Policy (2021), 3: e32. http://doi.org/10.1017/dap.2021.20.

[3] Tumber, Howard, Silvio Waisbord (eds). "The Routledge Companion to Media Disinformation and Populism". Routledge, London 2021, ISBN 9780367435769.

[4] Wardle, Claire. "Understanding information disorder. First Draft's Essential guide". First Draft, October 2019. https://firstdraftnews.org/wp-content/uploads/2019/10/Information_Disorder_Digital_AW.pdf?x14487.

[5] Wardle, Claire, Hossein Derakhshan. "Information Disorder. Toward an interdisciplinary framework for research and policymaking. Council of Europe report DGI(2017)09". Council of Europe, Strasbourg, 27 September 2017. https://rm.coe.int/information-disorder-toward-an-interdisciplinary-framework-for-researc/168076277c.

[6] Wardle, Claire, Hossein Derakhshan. "Information Disorder. Toward an interdisciplinary framework for research and policymaking. Council of Europe report DGI(2017)09". Council of Europe, Strasbourg, August 2018, 2nd revised edition. https://rm.coe.int/information-disorder-report-version-august-2018/16808c9c77.

[7] Wardle Claire, Hossein Derakhshan. Thinking about "information disorder": formats of misinformation, disinformation, and mal-information [In:] "Journalism, 'Fake News' & Disinformation. Handbook for Journalism Education and Training". Ireton Cherilyn and Julie Posetti (Eds). UNESCO, Paris, 2018, pp. 44–56. ISBN 978-92-3-100281-6.

[8] Wasserman, Herman (ed.). "Meeting the challenges of information disorder in the global south" [Research report]. 2022. https://idl-bnc-idrc.dspacedirect.org/bitstream/handle/10625/60954/IDL-60954.pdf.

Netography

[9] Kasprzyk, Rafał. INFOOPS (część 2/3): Dezinformacja jako taka, a jak nie taka to jaka https://www.cybsecurity.org/pl/infoops-czesc-2-3-dezinformacja-jako-taka-a-jak-nie-taka-to-jaka/.

[10] Key Internet Statistics in 2022 – The State of Cyberspace. https://www.internetadvisor.com/key-internet-statistics.

Aleksander Żołnierski and Jan Grzegorek

Migrants: Reactions to the Refugee Crisis 2021–2022

Contexts

The recent events related to the humanitarian crisis on the Polish-Belarusian border may be an interesting line of research in both sociology and economic anthropology. A wave of racism and Islamophobia flooded Europe. Poland was not isolated, but the scale of the problem here was far greater than in other countries. Only a small fraction of citizens stood up in defence of the immigrants (Buchowski, 2017). The Polish response, recorded already in the context of the 2015 refugee crisis, proved the most brutal and hostile in the European Union. Crossing borders was prohibited and the EU reprimanded Poland for the lack of solidarity. The government declared that it would accept refugees and in the official narrative of Law and Justice (PiS) that took power in Autumn 2015, Poland accepted "one million refugees from Ukraine" – Prime Minister Szydło used such wording in her official statement in the European Parliament in January 2016. The Ukrainian media responded with scepticism. They quoted the official statistics of the Polish government to state that in 2015, Poland granted refugee status to only two (sic!) Ukrainians. In addition, Radio Svoboda stated that economic migrants do go to Poland and in 2015 their number was a little more than 400,000 (which was the number of work permits issued). The government responded to the EU accusations by stating that the refugee acceptance process depends on whether the people applying for such status are Muslims. At present, more than two million Ukrainians work in Poland, most of them living in large cities. The PiS government makes no effective efforts to use the influx of workers and refugees from Ukraine and other countries to counter the depopulation of certain regions. However, the surveys conducted among Ukrainians in Poland showed that as many as 59 % of them would move to Germany upon the opening of that labour market (Toronto Star, 30.03.2019).

In 2015, PiS took power using the rhetoric of "Poland for Poles" – even if such slogans were not voiced officially, they structured the election rhetoric of the party of Kaczyński. The anti-immigrant rhetoric became a concern for the EU leaders. Prime Minister Morawiecki defined his purpose as not only to "re-industrialise" Poland but also, broader and more "ambitious", to rechristianise

Europe. The economy of Poland is one of those that benefited the most from the EU open market and openness to economic migration in the broader sense. The treasury received billions of euros paid by Polish workers. Facing shortages in the workforce, Poland used foreign workers, mainly from Ukraine and Russia, but also from Afghanistan, Belarus or Vietnam. Vice-minister of Investment and Development Chorąży, during a television debate before the local elections in October 2018, stated that "the influx of immigrants to our country must increase" and that "Poland must, and even if it does not want it, it should want it". His statement was not only not appreciated by the government of Morawiecki, but also decisively denied by Minister of Internal Affairs Brudziński, who remarked that it was not the "position of the government", and by Morawiecki himself, who said that the minister "clearly went too far in some of his statements" and dismissed Chorąży.

Nationalistic rhetoric, which is the foundation of the symbolic violence of the Polish government, uses in parallel the historical policy and the difficult Polish-Ukrainian relationships from the past. The combination of "national identity" built on the expression of brutal historical facts, but also legends, semi-truths, understatements and ordinary lies, while preventing expressions from the other side, is difficult to align with the anti-immigrant and racist policies towards non-Christians. In this context, the construction of the border wall is an important symbol (because this structure actually fails to serve any other purpose). The PiS government stated that accepting Muslim refugees may constitute a safety threat. These "modern" symptoms of racism correspond with the position of the Catholic church in Poland and a large portion of society. At present, the novelty complementing the "traditional" xenophobia, chauvinism and racism in the mainstream of the Polish church, namely the character of the Muslim terrorist, who only pretends to be a refugee, is a relatively new thread (Narkowicz, 2018). In the official narrative of the PiS government, Poland should help refugees "in situ", in Syria. The xenophobic anti-immigration and anti-refugee rhetoric set the tone for the public discourse. This permission introduced to the public space contributed to the overt expression of racist contents in the broadly understood Polish public discourse and society (Krzyżanowska, Krzyżanowski 2018).

To the rescue of the more or less covert chauvinism of PiS activists came the titles of the popular gutter press and state-subsidised newspapers connected with the authorities. Kaczyński, the leader of PiS, widely quoted also by the press that distanced itself from the PiS government actions and by the foreign press, stated at an election campaign gathering in Maków Mazowiecki "Different kinds of parasites, protozoans, that are not dangerous for the bodies of these people, may be dangerous here". At the PiS programme congress in Summer 2017, he elaborated on his thoughts by outlining the desired Polish response to the humanitarian disaster faced by the population of Syria due to the war supported by Russia: "We have the full moral right to say »no«" and he justified Poland's refusal to accept refugees for economic reasons "there is no reason for us to drastically lower our standard of living [...] it would be a social disaster, [...] it would jeopardise our ordinary everyday safety, [and] economic problems would appear along with the refugees" (Kim, 2017).

Fig. 1. Polish gutter press and populist politicians – one voice on refugees

The press also released (and still releases) disinformation texts on refugees and ordinary fake news – also the mainstream press, however they are relatively effectively verified.

Fig. 2. Disinformation and fake news on refugees are relatively effectively verified in mainstream press (here: Gazeta Prawna)

Xenophobia became a building block for the concept of identity, which consisted in focusing on some aspects of the Polish culture, while excluding a broader perspective on the world (Balogun, 2020). The public discourse begins to be dominated by a new narrative based on assigning new meanings to the existing nomenclature. In general, the shift of the language used in relation to immigrants involves all social groups, but its origins should be traced to the official narratives that brought PiS to power. In this process, language shift is evidenced by changes in categorisation. The media picked up, in response to the social need, the PiS narrative, thus producing ignorance by wilful sowing of doubts. After the 2015 refugee crisis, the word "uchodźca" (refugee) began to acquire increasingly negative connotations (Horolets et al. 2020). A refugee, newcomer from the areas of war, began to be called a "nachodźca" (unwelcome visitor) and increasingly associated with a security threat. The association between the notion of "refugee" and humanitarian matters weakened, while the proposition to "help in situ" gained social acceptance and the discussion on the moral meaning of that notion was overshadowed.

Short Chronology of the Humanitarian Crisis on the Belarus Border

The humanitarian crisis began in August 2021, the first reports indicating 8 August, when a group of 32 refugees from Afghanistan[1] crossed the border of Poland and were detained on the Polish-Belarusian border in Usnarz Górny. On 20 August, the government enacted a regulation to restrict border traffic even further and implement pushbacks. On 25 August, the European Court of Human Rights stipulated interim measures and ordered Poland to support refugees with food and shelter. In September (2.09), the President declared a state of emergency in the two voivodeships bordering Belarus. The decision completely prohibited access to the area for the media and support organisations. On 27 September, the European Court of Human Rights expanded the previous interim measures and extended their duration until further notice. Poland failed to comply with these obligations. One of the further steps of the Polish government was the so-called removal act (introduced on 26 October), and in early November the President signed the act on the construction of a six-metre high fence for more than PLN 1.6B (2 November). In the meantime, more refugees died at the border – on 5 December, the eleventh (official) casualty of the humanitarian crisis was recorded (a 38-year-old refugee woman from Kurdistan), while on the night of 6 to 7 December, a married couple from Iraq, pushed back to Belarus by the Polish Border Guard, reported that their 4-year-old daughter Eileen was missing. Still before Christmas, on 21 December 2021, the UN Office of the High Commissioner for Human Rights (again) called on Poland and Belarus to resolve the situation at the border in accordance with the laws binding both states.

The humanitarian crisis and border guard activity aroused media interest. The thousands of frozen and desperate migrants, against whom Polish forces were mobilised along the border, were the subject of reports by multiple media outlets, including the New York Times (Higgins, Pronczuk, Magdziarz, 2021). Finally, the Polish forces gathered along the border used water cannons and tear gas against the migrants. The Belarusian authorities were accused of "arming" the migrants with smoke grenades and weapons (police spokesperson Ciarka stated that the migrants were "pacified", because they had thrown stones at the

1 Russia, besides e.g. Pakistan, China and Iran, was one of the countries with good relations with the Taliban, who took power in Afghanistan in mid-August 2021. The Taliban signed an agreement with Russia for deliveries of gasoline, Diesel fuel, gas and wheat to Afghanistan.

Polish police). The use of tear gas was confirmed by the Ministry of Defence. Poland closed the border crossing in Kuźnica, so the narrative that the migrants wishing to apply for asylum were to cross the border at the formal crossing was not true.

The actions of the PiS government were condemned for the pushback conducted and for escalating the humanitarian crisis and, above all, for the failure to observe the law by preventing applications for asylum. Many migrants found themselves at the border with children, desperately trying to enter the EU in their escape from conflicts, poverty and instability not only in the Middle East but also in other regions (Gera, Litvinowa, 2021).

At the same time, most migrants planned to move to Germany and did not regard Poland as their final destination. It was not only the information on the differences in wealth between the two countries that influenced the decisions of the people at the border, but above all the knowledge of the opportunities and of how people of colour were treated in the two countries. National Public Radio (NPR) described the situation at the border from the Belarusian side, where a transfer facility for migrants was established. The facility hosted many families with children, most from Iraq, especially from Kurdistan, also from Syria, and some from African countries (Maynes, Block, 2021).

The Russian invasion in February 2022 made Poland face the challenge related to the migration of war refugees from Ukraine, but also in that case the border services took action that was racist in nature. The problem of racial segregation at the eastern border of Poland was addressed by USA Today. While Europe accepted the people escaping Ukraine, people of colour would be often sent back at the borders. In contrast to the migrants from Afghanistan, Syria, Iraq and Yemen, stopped at the border and subject to the pushback method, Poland was opened to people from Ukraine escaping war. However, not all refugees from war-torn Ukraine were treated in the same manner. Racial segregation took place already in Ukraine – reportedly, the Ukrainian police and security services refused Nigerians access to the buses and trains heading towards the Ukrainian-Polish border – as reported by the office of the president of Nigeria. Nigerians could also face similar treatment by Polish state officials, who refused them admission to Poland from Ukraine. Polish state officials tried to deny the accusations of racist treatment.

Already before the Russian invasion of Ukraine, record numbers of forced migrations were recorded worldwide – the UN Refugee Agency estimated that number at 84 million people. The UN estimated that more than four million people would flee Ukraine, which could be one of the greatest refugee crises in history. "The swift and coordinated UE response to the refugees from Ukraine is in sharp contrast with the EU response to the refugees from the conflicts in Afghanistan, Iraq, Palestine, Syria, Libya and Yemen," said Susan Akram, director of the International Human Rights Clinic, Boston University School of Law. "Despite many calls for Europe to launch the directive on temporary protection,[2] to grant legal status to the six million Syrians, who fled Syria from the beginning of that conflict, coordinates action to accept Syrian refugees was never taken" – she claimed. The policy of *helping in situ* resulted in thousands of deaths of refugees in camps in Lebanon, Turkey, Jordan, Libya, Egypt, Greece and Italy. Agreements were imposed with countries like Libya and Turkey, in which they were obliged to hold the refugees within their territories in exchange for more humanitarian aid funding from Europe. Putin's support for the Syrian dictator Assad and the Russian bombardments of the civilians in Syria did not influence the empathy and support for the refugees from that country. The American Arab and Middle Eastern Journalists Association condemned the racist migration policy. Segregation in treatment of the victims of war was condemned (Hauck, 2022).

Media and Propaganda

The development of communications methods using electronic means, in particular the development of social media, offers many solutions, driven by artificial intelligence algorithms, that on the one hand form peculiar information bubbles and on the other hand escalate the conflict by polarising social groups, the users of these media. The role of symbol increases, as a significant distinctive feature of the given group or tribe, while political divisions move beyond the election process. Some authors describe the loss of meaning of class voting and the increasing role of symbols as the decisive factor determining political choices. In 2015, such a symbol in Poland were the refugees from the war-torn Syria; today they are various groups that the PiS authorities chose to discriminate against, whereby the distinctive feature is gender. The authorities base that selection on current surveys and seem to address the social expectations of their voters. These surveys indicate that the media framing of these subjects excludes in the iterative process the persons that express views contrary to the prevailing view. These effects have a significant impact on the public debate, while aggravating the problem of racism. Research conducted in Poland confirmed empirically that hostile media presentation of migration changes the attitudes of respondents

2 The directive introduced in 2001 after the war in Yugoslavia.

regardless of their cognitive engagement. The purpose of the survey study was to check if contact with media information had an impact on attitudes (Wenzel, Żerkowska-Balas 2019).

Controlling the media is one of the objectives, which when achieved, increases the propaganda effectiveness of the governing party. In Poland, PiS took action to subjugate private media. Due to such actions, government propaganda, including racist and xenophobic contents, may effectively reach a wide audience. It should be emphasised that the regime in Russia is even more effective in subjugating the media. It launched Yappy, a local version of a video-sharing platform. The platform is owned by Gazprom Media. In addition, to take control over international social media platforms, Putin's regime enacted laws to oblige social platforms to open offices in Russia. It even created a controlled platform alternative to Facebook, VKontakte.

Description of the Sample for Sentiment Surveys

The sentiment surveys used as material for analysis were conducted in July–September 2022 on the sample of 158 selected profiles on Twitter (above all profiles of journalists, politicians and opinion-shaping representatives of political parties). From the set of tweets published on the selected profiles, two sets were formed: one published from 1 June 2021 to the day of the Russian invasion of Ukraine and another from the invasion day to the end of July 2022. Tweets were identified and collected with the character string of "migran" or "chodź" (as in migranci [migrants], imigranci [immigrants] and uchodźcy, nachodźcy [refugee, unwelcome visitor]), as well as all tweets with the character string of "immigrant" or "refuge" (to enable the analysis of tweet in English). In addition, all tweets that included the names of the inhabitants of the following countries were extracted and collected: Ukraine, Afghanistan, Syria, Lebanon, Iraq and Iran in Polish (e.g. Ukraińcy) and in English (e.g. Ukrainians), the selection of names in Polish was made following the rules of inflection of names. The application of the described criteria enabled the formation of the collection (T_0) with 16,417 documents. The collection will be divided into subsets: T_1 – tweets on refugees at the Belarusian border and T_2 – tweets on refugees from Ukraine. A working division into further subsets was prepared according to the country mentioned in the tweet content.[3]

3 In set T_0, the following subsets were identified (the values refer to the tweets in both Polish and English):

T_U – 8,517 tweets on the inhabitants of Ukraine

From the total number of tweets (T = 16,417), 14,214 were written in Polish and 2,203 in English. The largest number of tweets in the set refer to the inhabitants of Ukraine and to refugees without mentioning their country of origin and almost all tweets referring to the inhabitants of Ukraine were published from February to June 2022. For further analysis, the sets were divided into two subsets, which was the final division of the analysis: the first part included the tweets published before 24 February 2022, i.e. before the Russian invasion of Ukraine, and the second part included the tweets published after that date.

The sets prepared in this manner were subject to stemming, consisting in the removal of the inflection suffix and leaving only the word stem, and lemmatisation, grouping together of the different word forms to allow statistical analysis of words identified by lemmata. Recurring entries were removed from the tweet content and the text was converted to lower case letters; all characters not constituting words were replaced with spaces and then all numbers and white spaces were removed from the text.

Results of the Study

The results of study indicate that the sentiments and emotions connected with refugees from territories threatened by armed conflict differ. The dependency between the polarisation of sentiments, but also their intensity, results from the country of origin of the migrants. Even selected accounts on Twitter include a significant difference of emotions depending on the perceived skin colour and religion of the people attempting to cross the eastern border of Poland. The preparation of the set of the most common words in tweets before and after the Russian invasion of Ukraine, defines the difference of notions and meanings used in the narratives related to each of the two analysed groups of refugees. This difference is well-illustrated by the cloud of tags consisting of the most common words in comments before and after the invasion.

T_A – 503 tweets on the inhabitants of Afghanistan
T_S – 186 tweets on the inhabitants of Syria
T_L – 14 tweets on the inhabitants of Lebanon
T_{IR} – 195 tweets on the inhabitants of Iraq
T_{IRN} – 60 tweets on the inhabitants of Iran
T_N – 7,124 tweets with no mention of inhabitants of the above countries

Fig. 3. Cloud of tags before the invasion

The tweets before the invasion were dominated by terms connected with claims of the illegal nature of the attempts to cross the border. At the same time, the commentators emphasised the matter of the irresponsibility of the parents of the children in the strip of land between Belarus and Poland, who were subject to pushback. In the dominant narrative after the invasion a definite majority of entries related to the inhabitants fleeing the territories of Ukraine contained terms indicating a positive attitude of the commentators toward Ukrainians and satisfaction with the willingness to help the refugees.

The negatively marked comments were directed towards Russians, while the Ukrainians themselves enjoyed positive sentiments. The comments on the situation after the invasion are illustrated in the following figure. Both cloud tags were subject to the "methods" that emphasised the size of the fonts used $\sqrt{n}$, where n meant the number of keywords searched.

Fig. 4. Cloud of tags after the invasion

Therefore, the described difference in the narratives related to war migrations is decisively clearer for anonymous accounts and the media that do not moderate comments posted online (or only interfere with the content to a small extent). An example might be the texts published on gazeta.pl. The lack of moderation of statements marked with decisively negative sentiments with strongly polarised emotions is particularly visible there.

The analysis of emotions in the studied set of tweets involved the identification and marking of the terms indicating the instinctive and spontaneous response of their authors. Twitter users described the phenomena that were the stimuli for specific reactions, which proved to be mainly negative in the first period studied. The stimuli (events, people, situations – refugees, migrants, Ukrainians etc.) could be both an external stimulus (in the form of media information, both in traditional media and in digital or information heard during personal relations of the author with the environment) and an internal stimulus, based on previous experiences. Such stimulus did not necessarily need to reflect an actual experience, it could be a result of the impact of imaginations or thoughts or imagined memories or an event created in the imagination. Within this meaning, the sentiments connected with migrants at the Belarusian border were formed analogically to the Polish "antisemitism without Jews". The sentiments, unlike emotions, result from psychical attitude, which could have its origins in the national imaginary exploited in recent years by the mainstream media and used to create the national identity resisting the real historical experience of the commentators. The sentiments expressed are therefore an aggregate of these emotions. Sentiments express subjective experience. Nonetheless, they appear as the identified description of the condition resulting from emotions, somehow built on real experience and thus expressed. Sentiments, unlike the emotions themselves, are a conscious response to a stimulus.

The difference in the emotions and trend shift in relation to before and after the invasion, expressed in the studied set of tweets, is shown Fig. 5. The trend shift connected with the influx of migrants from Ukraine is clearly noticeable. The analysis of differences in the polarisation of sentiments in the studied set of tweets indicates a clear trend shift in the studied period. Fig. 5 shows the difference between the positive and negative polarised statements of the users of Twitter. The dominance of negatively marked statements (late December 2021 and early January 2022) was broken after the invasion of Ukraine. The trend line describing the change of attitude in the comments (marked with blue dotted line) begins to change its course and the prognosis indicates that migrants from Ukraine may enjoy a friendly welcome in Poland.

Fig. 5. Positive and negative sentiments before and after the Russian invasion of Ukraine (07.2021–06.2022)

[Sentiments related to refugees from conflict zones (07/2021–06/2022). Red: negative trend; green: positive trend; black line: difference; blue dotted line: trend – change in sentiments]

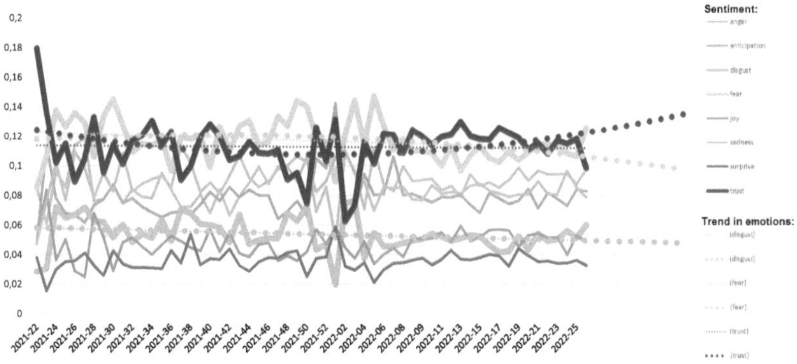

Fig. 6. Emotions before and after the Russian invasion of Ukraine

The analysis of the emotions connected with the markedness of the sentiments expressed in the analysed tweets is interesting. Disgust towards refugees, fear of them and anger build negative attitudes and sentiments to the greatest extent.

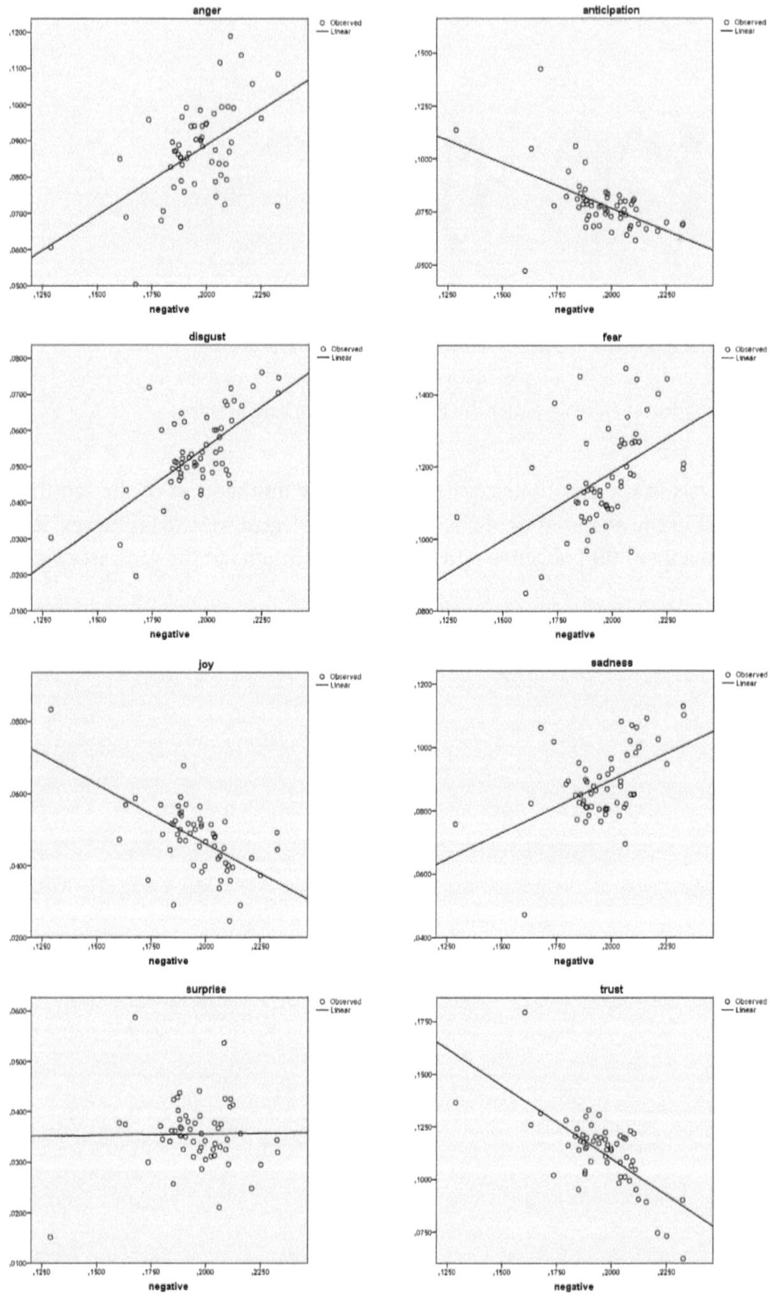

Fig. 7. Negative sentiments and emotions

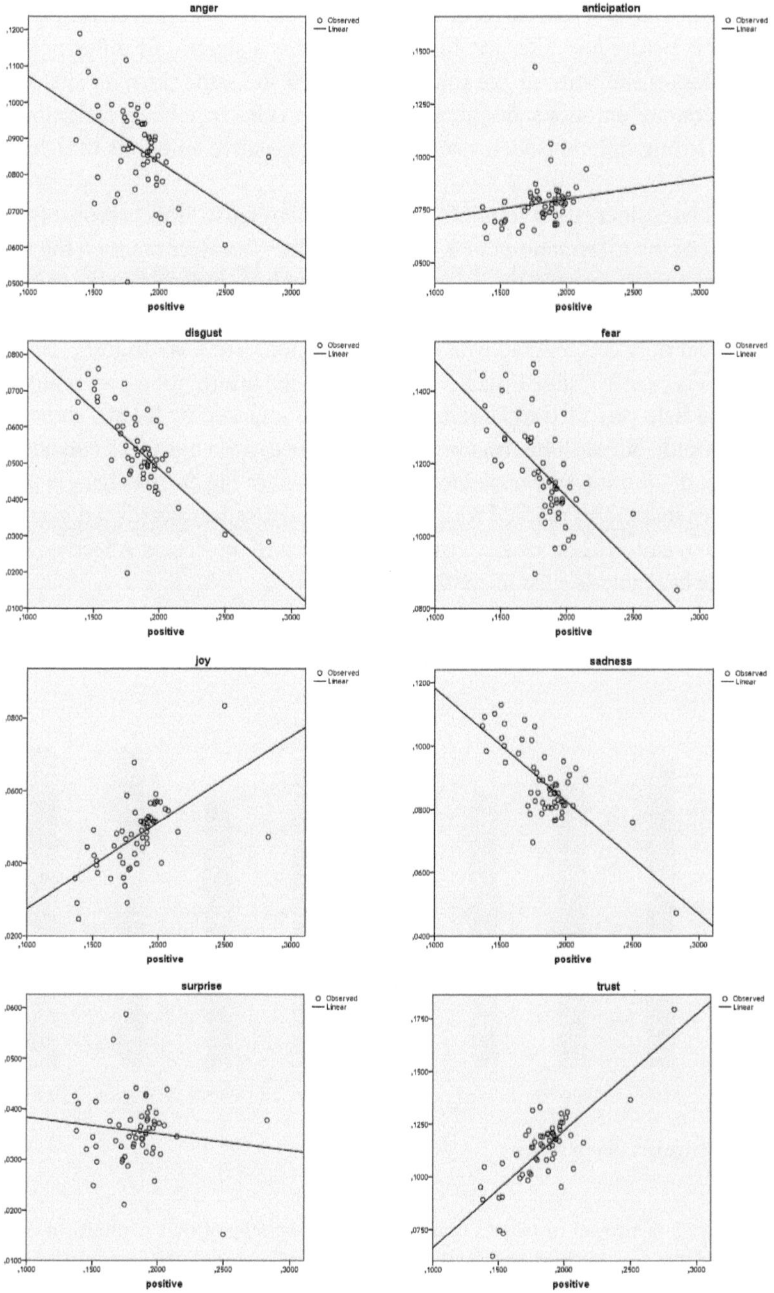

Fig. 8. Positive sentiments and emotions

The juxtaposition of the totals of the simple number of negative and positive sentiments before and after the invasion indicates a significant difference and domination of the latter in the studied tweets. At the same time, an analogous juxtaposition of emotions, despite the percentage difference being insignificant, shows a strong difference between positive and negative emotions in dynamic terms.

The greatest increase is found for "joy" and then "trust". This is probably also the result of the information block imposed by the PiS government on the news describing the situation on the Belarusian border, while emphasising the attitudes of Poles and the willingness to help Ukrainians. At the same time, the decrease in the number of tweets marked with negative emotions, such as "disgust", "fear" or "sadness", is clearly visible It seems that not only the information policy but also the actual help provided to Ukraine when it was attacked by Russia, shows the actual attitude of Polish Internet users towards the eastern border of Poland. The dynamics of emotions expressed in tweets is shown in Fig. 9. The changes in the number of statements marked with fear, anger and disgust towards refugees that occurred in early 2022 is clearly visible. Fig. 9 is based on Nencki Affective Word List made available as a list of 2,902 Polish words.

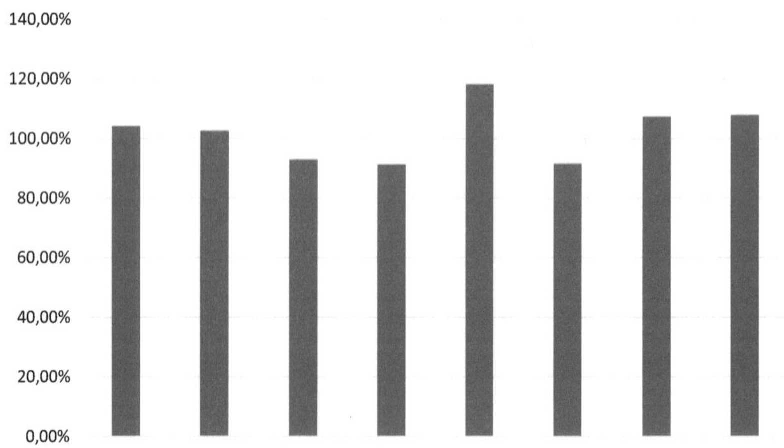

Fig. 9. Dynamics of emotions

The analysis of the set of tweets using the SPSS package allows evaluation of the quality of the model built using the frequency of occurrence of emotions and polarisation of sentiments. The results of the analysis indicate that the model is most suitable for two emotions: "fear" and "sadness". In these cases, the measure of the quality of the suitability of the model, the determination factor, is the highest. The quality of the model is illustrated in successive figures, which show the individual types of emotion in terms of the monitored variables and indicating the degree of suitability.

Conclusion

The attitudes related to the humanitarian crisis at the Polish-Belarusian border and the influx of migrants from Ukraine after the Russian invasion of that country are an interesting area of research including many disciplines, from security studies to economic sciences, to sociology and political sciences or economic anthropology.

The comments related to refugees had a clearly negative meaning and Twitter was one of the many media dominated by posts filled with fear, aversion and anger towards refugees with a perceived darker skin colour. The narrative directed against accepting refugees was the set of economic "arguments", but it seems that racism was an important factor determining the aversion. The situation changed after the Russian invasion of Ukraine, when great numbers of Ukrainians began to cross the Polish border. The analysed comments on Twitter shifted, while the sympathy towards Ukraine and its inhabitants overcame the fear and unwillingness to accept migrants. The terms "help" and "accept" appeared in the narrative and previous bias, although still present, stopped dominating the studied tweets. The posts expressing "joy", followed by "trust", while emphasising the positive and friendly attitudes of Poles towards refugees from Ukraine, became prominent in the Internet narrative.

Does the wave of aversion towards migrants, its dynamics and course of changes in the sentiments show symptoms of the behaviours that constitute the statistics of racism and xenophobia in Poland? Considering the media in which moderation by platform administrators is scarcely applied, are there any additional factors behind the wave of hate towards the migrants? Regardless of the answers to these two questions, one thing is certain – the immense support for the most extreme opinions and statements cannot be denied. In addition, if "fear" is considered as the factor that to the greatest extent explains the negative sentiments identified in the content of tweets, it proves that effective management of that feeling can be (and perhaps is) essential to the achievement of the long- and medium-term political objectives, as well as the ongoing biopolitical activities of the current party governing Poland.

From the point of view of stability of the social and economic environment and questions of internal security, it should be emphasised that the accumulation and reproduction of xenophobic and nationalistic trends in public space are contrary to the principle of solidarity and European identity, thus driving the activities that could be described as all kinds of -exits. The increasing deficits in social capital in Poland, connected with the explosion of symbolic violence, may soon lead to Polexit, of course in de facto, not legal terms. The economic effects of these processes are coming soon and the increasing deficit in social capital is expressed both in the increasing transaction costs and low economic innovation, in particular in activities that require cooperation and trust.

The escalation of the conflict at the Polish-Belarusian border raised awareness of the potential problem of migration of a large number of Ukrainian citizens if Russia decided to invade. At the national level, action was taken to use this migration to attempt to change the image of Poland in the EU (after the definite image crisis caused by systemic violations of the EU law by the PiS government – the crisis related to the so-called Disciplinary Chamber, the Constitutional Tribunal, the conflict with Czechia related to the Polish open-pit mine in Turów or the humanitarian disaster caused by the racist migration policy). In hindsight, it could be said that the response of Polish services contributed to the negative perception of Poland in the countries of the West.

In addition, the humanitarian crisis at the Belarusian border defined new standards in relations with migrants from non-Christian countries, as thousands of refugees were trapped between troops of the Belarusian and Polish border guards. The treatment of those, who managed to enter Poland, subject to organised violence and planned pushbacks, as they were captured and expelled into no-man's land, which one might refer to as *Sonderbehandlung*.

To conclude, the conducted research sheds a new light on many processes that are fundamental to government activities that are popular among PiS supporters. In addition, the research is a potential point of departure to further exploration of these phenomena, as well as asking further research questions or building scenarios for strategy and politics. The key question remains: what would be the potential consequences of actions other than those taken by the PiS government in relation to the humanitarian crisis at the Belarusian border? Would opening the border for refugees from other war-torn territories, in which Russia played a crucial role, be a signal for Putin that Poland, the European Union and the West would provide that massive support for Ukraine? If the dictator in Moscow had known that war refugees and the nations he decided to conquer would find support in Europe, would he have invaded Ukraine? Apart from these key questions, the research provides grounds for the following further hypotheses:

1. The humanitarian crisis at the Polish-Belarusian border was undoubtedly planned in Moscow and executed by Lukashenko's services. To what extent

did the crisis affect the image of the party governing Poland in foreign (including Russian) media?
2. Was a significant number of comments (above all in non-moderated media) with hate towards refugees created by Russian trolls or by the Polish far-right?
3. The humanitarian crisis and the involvement of Belarusian and Russian services could have been a dress rehearsal before the expected influx of migrants from Ukraine – if Russia planned the invasion much earlier, would the anti-immigration attitude of Poland benefit Moscow?
4. Assuming that the third hypothesis is true, could it be said that the Russians underestimated the degree and scope of attitudes rooted in racism and xenophobia in Poland?
5. Anti-immigration sentiments fuel political support for PiS – could Kaczyński's party have had an interest in escalating the crisis?
6. The Russian invasion of Ukraine and the outbreak of war influenced the change of the official attitude of the PiS government toward refugees (although positive sentiments were only directed towards immigrants from Ukraine). Did that shift in the narrative change the perception of Poland and the PiS government in foreign media?

References

Balogun, Bolaji. "Race and Racism in Poland: Theorising and Contextualising 'Polish-Centrism'". Sociological Review 68, no. 6 (November 2020): 1196–1211. http://doi.org/10.1177/0038026120928883.

Buchowski, Michał. "A New Tide of Racism, Xenophobia, and Islamophobia in Europe: Polish Anthropologists Swim Against the Current". American Anthropologist 119, no. 3 (September 2017): 519–523. http://doi.org/10.1111/aman.12915.

Maynes, Charles, and Melissa Block. "Hundreds of Migrants Remain Trapped at the Belarus-Poland Border". Morning Edition (NPR), 22 December 2021. https://search.ebscohost.com/login.aspx?direct=true&db=nfh&AN=6XN2021122207&lang=pl&site=ehost-live.

Hauck, Grace, and USA TODAY. "Not All Refugees given a Welcome". USA Today. Accessed 1 July 2022. https://search.ebscohost.com/login.aspx?direct=true&db=f5h&AN=J0E005571873822&lang=pl&site=ehost-live.

Higgins, Andrew, Monika Pronczuk, Anatol Magdziarz, and James Hill. "Migrant Crisis at Poland's Border Eases Pressure on Government". New York Times, 25 November 2021. https://search.ebscohost.com/login.aspx?direct=true&db=f5h&AN=153766237&lang=pl&site=ehost-live.

Horolets, Anna, Adriana Mica, Mikołaj Pawlak, and Paweł Kubicki. "Ignorance as an Outcome of Categorizations: The 'Refugees' in the Polish Academic Discourse before and after the 2015 Refugee Crisis". East European Politics & Societies 34, no. 3 (August 2020): 730–751. http://doi.org/10.1177/0888325419891204.

Kim, Renata. "Niemoralne słowa Kaczyńskiego o uchodźcach", Newsweek 2 lipca 2017, aktualizacja 3 lipca 2017 08:11, https://www.newsweek.pl/opinie/jaroslaw-kaczynski-niemoralne-slowa-o-uchodzcach/206vq4l.

Krzyżanowska, Natalia, and Michał Krzyżanowski. "'Crisis' and Migration in Poland: Discursive Shifts, Anti-Pluralism and the Politicisation of Exclusion". Sociology 52, no. 3 (June 2018): 612–618. http://doi.org/10.1177/0038038518757952.

Marc Santora The New York Times. "Poland Bashes Immigrants, but Quietly Takes Christian Ones". Toronto Star (Canada), March 30, 2019. https://search.ebscohost.com/login.aspx?direct=true&db=nfh&AN=6FPTS2019033053922810&lang=pl&site=ehost-live.

Narkowicz, Kasia. "'Refugees Not Welcome Here': State, Church and Civil Society Responses to the Refugee Crisis in Poland". International Journal of Politics, Culture & Society 31, no. 4 (December 2018): 357–373. http://doi.org/10.1007/s10767-018-9287-9.

Vanessa Gera And Daria Litvinova The Associated Press. "Poland Uses Water Cannons against Migrants at Border". Toronto Star (Canada), November 17, 2021. https://search.ebscohost.com/login.aspx?direct=true&db=nfh&AN=6FPTS20211117628109 41&lang=pl&site=ehost-live.

Wenzel, Michał, and Marta Żerkowska-Balas. "Framing Effect of Media Portrayal of Migrants to the European Union: A Survey Experiment in Poland". East European Politics & Societies 33, no. 1 (February 2019): 44–65. http://doi.org/10.1177/0888325418777058.

Iuliia Arkhipova

Dissemination of Russian Propaganda via Social Media: The Case of the Vinnytsia Missile Strikes

Introduction

Since the beginning of the full-scale Russian invasion of Ukraine, the Russian propaganda machine has been handling new challenges daily. It has to make Russians believe that Ukrainians are their enemies, that the Russian government is trustworthy, and that the Russian military is running a "special military operation" targeting only "Ukrainian Nazis" and without doing any harm to civilians.

According to Deloitte [2], 89 % of Russians in 2021 watched TV at least once in the previous two weeks, 81 % consumed media content online, and 90 % used social media. 82 % used WhatsApp, 52 % Telegram, 49 % Viber. According to a Russian media agency Mediascope [4], about 40 % of Russians visit VKontakte daily and 35 % Telegram (since February 2022), about 28 % visit TikTok and 17 % Odnoklassniki.

Telegram is a messenger with a huge network of public pages. Here messages are first discussed in the network of government-related bloggers (including TV "talking heads") and anonymous public pages. Most politically active celebrities choose this messenger as their primary platform for communication with their audience. Bloggers (including those working in the warzone) and anonymous channels' messages can be re-posted by these celebrities, so they also choose to actively use this platform. Telegram's channels are rarely censored, so use of derogatory language and dehumanising methods is widespread here.

VKontakte is the major social network in Russia. Here the messages are disseminated for a wider audience through networks of public pages, groups and bloggers. An important role here is played by regional and municipal level public pages (municipalities, regional departments of political parties, even municipal libraries and kindergartens). Private commercial local publics such as "Podslushano (Overheard) in <name of a city or town>" or "Tipichny (Typical) <name of a city or town>" also sometimes post such propaganda messages. During the COVID-19 pandemic the system called "Posev (Seeding)" was introduced [3]. It consisted of "fabrics of content" from which local publics were supposed to take news for dissemination. For example, more than 100 pages of municipalities and municipal entities, including kindergartens and libraries, posted the same text on July 6 (they can be found during searching a phrase "мы расскажем о старшем лейтенанте Алексее Гришине. Он лично сбил штурмовик СУ-25" ["we will talk about senior lieutenant Alexei Grishin. He personally shot down an SU-25 attack aircraft"] on VKontakte). In the same manner a video about a sapper was disseminated on VKontakte and shared more than 200 times (they can be found during searching a phrase "Вячеслав рассказал, какие смертельно опасные мины расставляют националисты, и почему их не всегда сразу можно обнаружить" ["Vyacheslav told what deadly mines are placed by nationalists, and why they are not always immediately detectable"]).

This pilot research was conducted on a small dataset (concerning one event) without using machine data processing. Its aim was to understand more about the dissemination of propaganda in Russian social media and to work on hypotheses for further research.

The Case of the Vinnytsia Missile Strikes

On July 14, 2022, at 9:45 CET three Kalibr rockets hit an office building, the House of Officers, residential buildings and a parking lot in the centre of Vinnytsia. In the result of the shelling 23 people were dead [1], three of them were children and at least 20 of them were civilians, including staff and patients of a private medical clinic [5]. Three minutes later, at 9:48, Russian propaganda channels started to publish news about explosions in Vinnytsia.

For this research, 927 Telegram posts and 379 VKontakte posts mentioning the word "Vinnytsia" (in Russian "Винница" or "винницкий" with different endings) were collected. Both Telegram and VKontakte channels stopped posting messages with these words a week after the event, on July 21. On Telegram it is possible to search for particular phrases only in the channels selected, so the search was conducted on 342 public channels, each having more than 10,000 subscribers and having more than 50,000,000 subscribers in total (average number of subscribers is about 150,000). These channels were identified using the "snowball method" and starting with accounts of people affiliated with RT (Russia Today). On VKontakte it is possible to search particular phrases among all the posts published by users, so in order to narrow down the number of messages only those which got more than 100 "Like" reactions were included in the database.

In the collected posts at least 45 repeating phrases were identified. As the research was conducted without using machine data processing, these phrases (later called "messages") were identified due to unusual vocabulary or particular word combinations. For instance, these were surnames of killed during the attack Ukrainian military men ("Бурдыко", "Макарчук") or phrases such as "at the moment of the meeting of buyers and sellers of Western weapons" (in Russian "в момент встречи покупателей и продавцов западного оружия"). The lifespan of such a message (from the first mention to the last one) amounts from 1 to 138 hours, 28 hours on average (see Fig. 1).

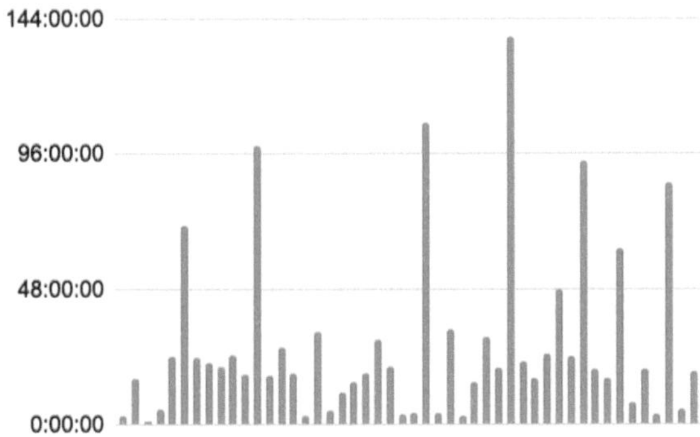

Fig. 1. Lifespan of a message (y = hours)

Both Telegram and VKontakte channels and pages lack unique content, most of which is produced by news agencies, opinion leaders, and "war reporters". Thus, the administrators of the channels rewrite or just copy and paste another's content in order to create their own. In most cases messages are disseminated in a top-down manner: first they appear in popular channels and then they are re-written or just copied and pasted in less popular ones. For instance, the message "The missile directly hit the east wing" ("Ракета ударила четко в восточное крыло", see Fig. 2) was posted in three popular Telegram channels and got 258,500, 387,800 and 300,800 views there, and then was posted in other channels 18 times, gaining 20,000 views on average.

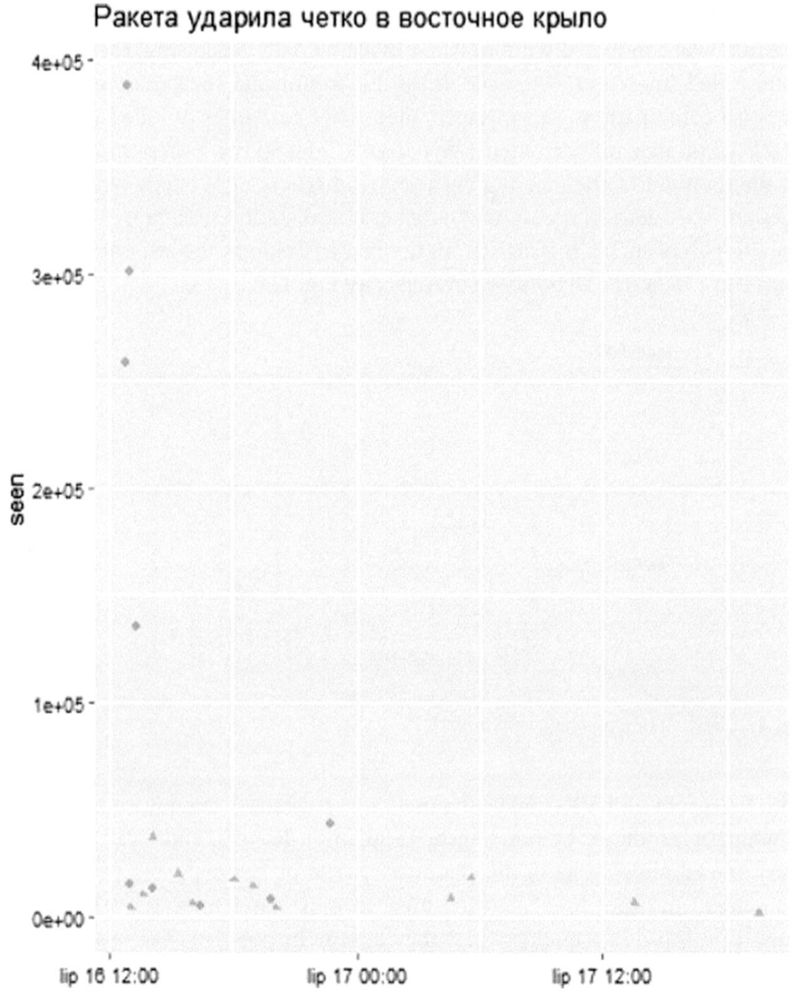

Fig. 2. Dissemination of the message "Ракета ударила четко в восточное крыло" (x = date and time of publication, y = number of times the message was viewed; circles represent Telegram posts, triangles represent VKontakte posts)

There is also a group of messages, produced in less popular channels and then re-posted in more popular ones. The message "damn orcs, what are they doing to innocent Ukrainian elves" ("проклятые орки, что творят с невинными украинскими эльфами", see Fig. 3) was first posted on VKontakte and got 8,200 views there, but being copied and published by a popular account on Telegram got 95,900 views.

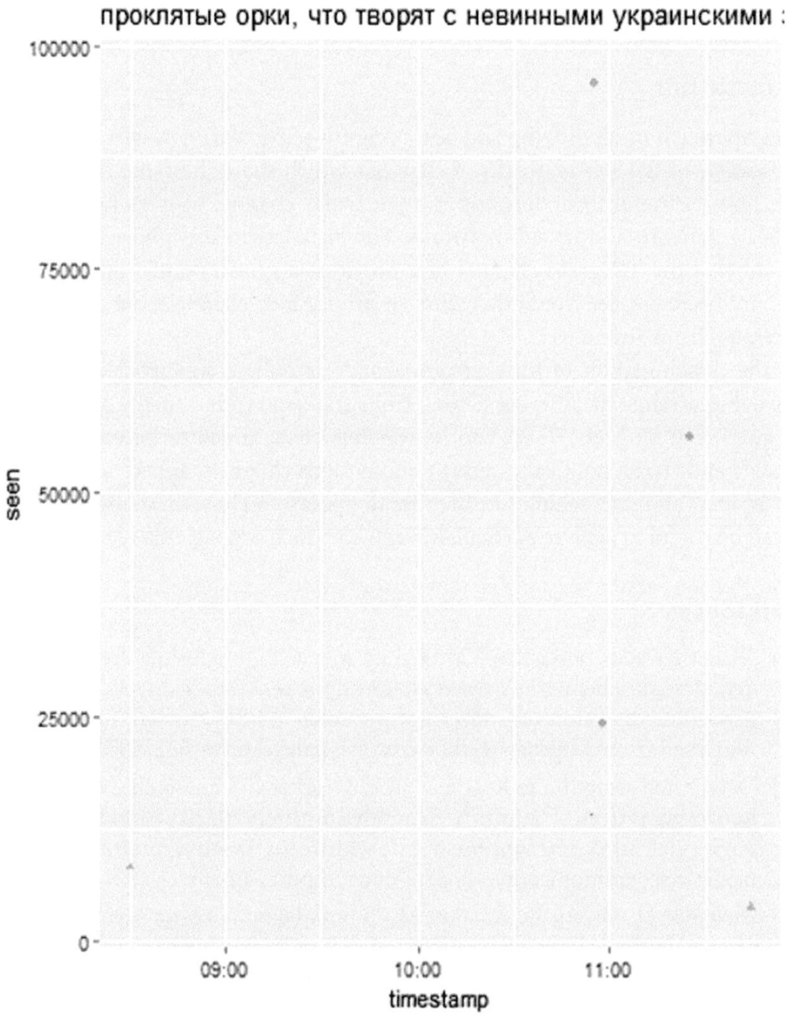

Fig. 3. Dissemination of the message "проклятые орки, что творят с невинными украинскими эльфами" (x = date and time of publication, y = number of times the message was viewed; circles represent Telegram posts, triangles represent VKontakte posts)

In most cases the messages were first published on Telegram, and later appeared on VKontakte pages, but the opposite was also possible. This could be explained perhaps by the popularity of the Telegram messenger among the opinion makers combined with VKontakte.

Conclusion

This approach of identifying and searching for particular messages also allows indication of the sub-networks of channels which share the same messages. It can also be useful to understand if a particular channel belongs to the propaganda network and its sub-networks. For instance, in this pilot study it was found that the Telegram channel of a pro-Russian media outlet Strana.ua [6] did not use messages from other Russian propaganda channels, but others took messages from Strana.ua.

The dissemination of Russian propaganda is quite a sophisticated process, involving a range of different actors, from state-sponsored media such as RT, to grass-root bloggers. These can be war reporters, aimed to become opinion leaders and start a political career, or anonymous channels, aimed to gain more subscribers and start selling ads. In order to understand how this network works, it can be useful to analyse particular events to which propaganda reacts.

References

[1] "Raketnyj udar po centru Vinnicy: ne menee 23 pogibshih, sredi kotoryh deti, desjatki ranenyh" ["Rocket attack on the center of Vinnytsia: at least 23 dead, including children, dozens of wounded"]. BBC Russian, 14 July 2022. Retrieved from: https://www.bbc.com/russian/features-62155826.

[2] "Media consumption in Russia – 2021". Deloitte CIS Research Centre. 2021. Retrieved from: https://web.archive.org/web/20220308011123/https://www2.deloitte.com/content/dam/Deloitte/ru/Documents/technology-media-telecommunications/media-consumption-in-russia-2021-en.pdf.

[3] Guskova, D., Shvets, D., Maglov, M. "S popravkoj na koronavirus. Tenevaja propagandistskaja set' ispol'zuet pandemiju dlja prodvizhenija izmenenij v Konstituciju" ["Coronavirus adjusted. Shadow propaganda network is using the pandemic to push constitutional changes"]. Mediazona, 21 April 2020. Retrieved from: https://zona.media/article/2020/04/21/videopopravka.

[4] "Auditorija Social'nyh Setej. Analiticheskij Otchet" ["Audience of Social Networks. Analytical Report"]. Mediascope, 2022. Retrieved from: https://mediascope.net/upload/iblock/de3/Social_media_Mediascope_18042022.pdf.

[5] "Raketnyj udar po Vinnice" ["Rocket strike on Vinnytsia"]. Wikipedia, 2022. Retrieved from: https://ru.wikipedia.org/wiki/Ракетный_удар_по_Виннице#Жертвы.

[6] Zamkovoi, O. "Strana.ua or Strana.ru? How a Sanctioned Site Disseminates Banned Kremlin Messages in Ukraine". Stopfake, 2021. Retrieved from: https://www.stopfake.org/en/strana-ua-or-strana-ru-how-a-sanctioned-site-disseminates-banned-kremlin-messages-in-ukraine/.

Halyna Synorub

Ways of Forming Informational Resilience of the Audience Under Wartime Conditions

Introduction

The invasion by the Russian Federation of Ukraine has opened not only a military front but also an information one, which envisages countering an intensive and negative information-psychological influence on individuals and the society in general. Acquiring the skills of evaluating and interpreting media information, recognising manipulative and propaganda content, preventing oneself from being drowned in the sea of disinformation at a time when Russian aggression has intensified, avoiding the euphory of sensation-linked addiction in the time of emotional challenges and fakes is the main task for Ukrainians, who have to achieve this in the most complicated circumstances ever. These days, the most refined information weapons are used to seek sensitive topics, shape opposing opinions within the society, find the people who share them, and then force the two camps into a confrontation. All-embracing information campaigns, large-scale projects and pinpoint insertion of fake news are all systemic developments of large complexes of propaganda narratives that are produced on various information platforms and "hook" the audience successfully. That is why it is most important to develop mechanisms of informational resilience that would enable us to stabilise and protect the audience from destructive information influences, and therefore secure victory over the enemy. This determines the relevance of the present research.

Informational resilience during the war has become the focus of attention for many researchers including I. Bachurin, N. Voitovych, V. Zhadko, O. Kuchyk, O. Klymenko, P. Kulias, O. Markiv, Yu. Poltavets, G. Pocheptsov, O. Pelian, O. Taranenko, O. Kharytonenko, O. Kharchuk, S. Shevchuk and others.

Under wartime conditions, the leading component of informational resilience is the psychological element, which is characterised not only by the ability to withstand the difficulties of life, the pressure of unfavourable circumstances while maintaining the capacity to work and the ability to acquire new experience, but also by flexibility, and the ability to analyse and comprehend information.

The problem of informational and psychological resilience is explored in the works by Ukrainian (Ya. Aminieva, T. Andreieva, I. Arshava, E. Nosenko, V. Beliumov, D. Zasiekina, Z. Koval, M. Kostenko, T. Mazur, O. Romanchuk, Kh. Shyshkina and others) and international (B. Bernard, Suzanne Kobasa, Salvador Maddi) scientists.

Under the current wartime conditions, human beings, society and the state are subject to strong informational and psychological influences. The problem of forming and expressing such qualities and abilities in the audience that combine balance, resilience, stability and resistibility, is growing more and more relevant [3]. The topic of informational resilience of the audience calls for thorough research since it is of critical importance in the period of the Russian-Ukrainian war.

The purpose of this study is to identify top priority levers in the informational influence of media environment on the audience in the war period, and to outline the main mechanisms that form resilience to disinformation.

Presentation of the Core Material

With the beginning of the Russian invasion, considerable changes have taken place in the Ukrainian media environment [2] (Fig. 1).

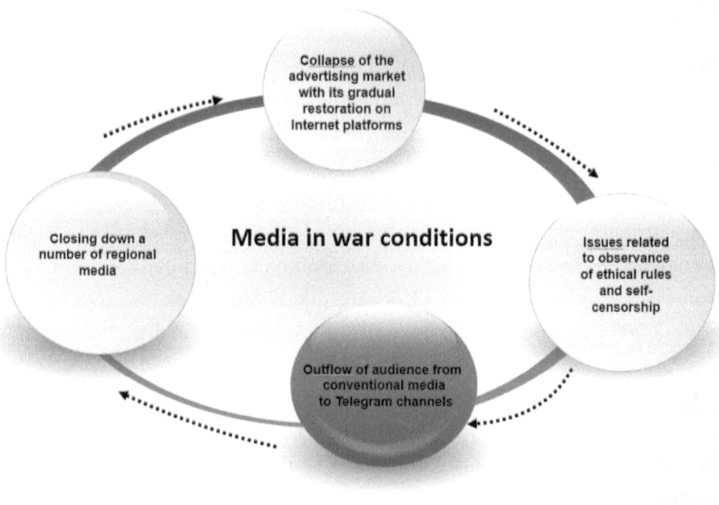

Fig. 1. Changes in the media environment during the war

Such transformations have changed media consumption drastically. In this process, a particular role belongs to the channels through which the audience receives information. According to the survey by the Kyiv International Institute of Sociology conducted on request of the OPORA Civil Network [5], the most popular source of information in 2022 was social media, as 76.6 % of Ukrainians used them to access news. Television ranks second, with 66.7 % of respondents. The third position in this rating belongs to the Internet (other than social media) as reported by 61.2 % of users. Radio is listened to by approximately 28.4 % of Ukrainian citizens, and only 15.7 % of the respondents read printed media (Fig. 2).

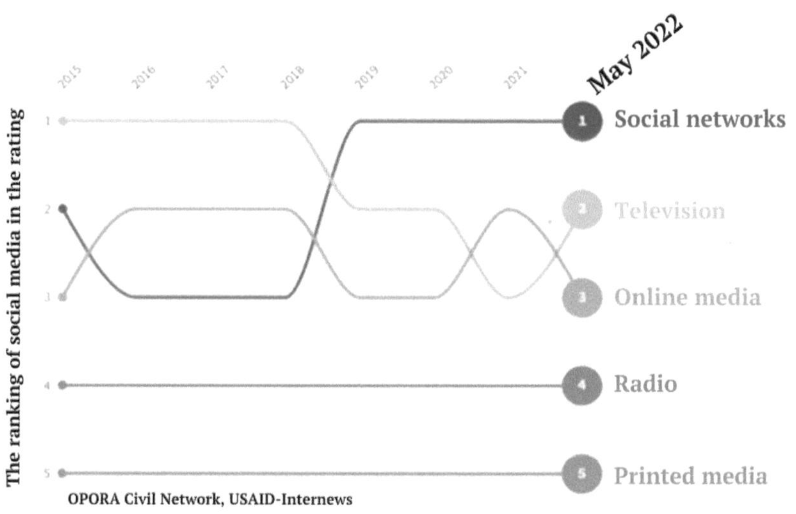

Fig. 2. Changes in media consumption in 2015–2022

Since 76.6 % of the citizens prefer social media as the primary source of information, it is worth exploring which of them enjoy the highest demand. For instance, 66 % of the respondents choose Telegram, 61 % – YouTube, and 58 % – Facebook (Fig. 3).

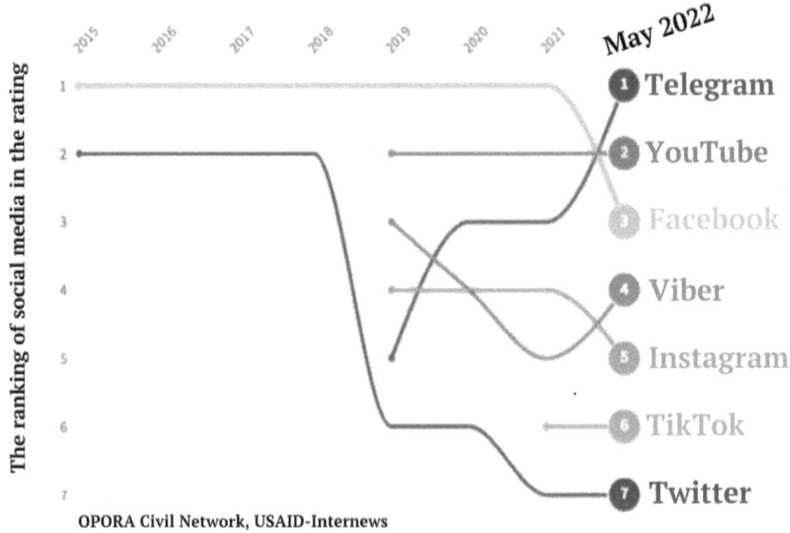

Fig. 3. The rating of most popular social media in Ukraine

It should be emphasised that during the war Telegram is one of the platforms that reports events as they occur in real time. It also frequently becomes an outlet for spreading fake news. This happens because there are numerous anonymous Telegram channels whose owners' identities are absolutely unknown. The specific feature of this platform is that it does not have a dedicated news feed, but contains a different mechanism instead – internal distribution channels, known as networks, that can be full of manipulative, propagandist, or fake messages. The naturally determined issue of protection against negative media influence, by way of forming informational resilience of both individual citizens and the entire society, has become utterly relevant.

Romanchuk views the notion of "resilience" in wartime conditions as a multifaceted personal quality which is expressed in feeling the values and content ("truth is with us, god is with us, victory is with us", "We are in our land, and we are fighting for our freedom and our future"), in the energy of light and love ("Warriors of light", "Angels of the AFU"), in the power drawn from unity of the people and their mutual support ("We are proud to be Ukrainian!"), in making a choice ("Our task and our choice: to hold out, win, and revive!"), in feeling hopeful ("We will win!"), in rational thinking ("The darkness will retreat. The dawn is coming. Hold the line…"), in steadfastness and infallibility ("We know our mistakes! We're working on them!"), and in the skill to regulate one's emotions. This is a disciplined rhythm of taking care of oneself and recovery, which is considerably influenced by the informational environment [8].

Melnyk provides an explanation of information resilience as development of not only adequate mechanisms to combat explicitly aggressive information influences, but also of mass-level metanarratives based on ethnic-national axiological component [6].

According to Koval, informational resilience consists in applying a complex of methods to wage information warfare and counter similar actions, in taking into account sociological and psychological aspects of online communication, and in monitoring activities of virtual community members [3].

We believe that informational resilience under wartime conditions is a security-related phenomenon which is achieved by elaborating strategies of combating disinformation and manipulative media content both on the level of the mass audience and government control, which will, in its turn, ensure strengthening the information frontline. The complex of informational-psychological influences shapes the sociocultural environment, so the main task of the state consists in ensuring support and providing the system of informational-psychological resilience, as well as securing information security. Under wartime conditions, with the increased level of real threats to people's lives, fear, desperation and uncertainty, the audience faces the task of identifying the information that exerts a negative influence on mass consciousness, national and cultural identity, and therefore facilitates the erosion of national values. It raises the issue of effectively ensuring information security in the context of theoretical and practical foundations of conducting information-psychological operations for achieving political, informational and other advantages over the enemy. The work to improve the ability of both the mass audience and media professionals to counter harmful information influences through upgrading skills of critical consumption of information, and overcoming cognitive prejudices, with account of emotional vulnerability of Ukrainians, should be both systemic and intensive during the Russian-Ukrainian war.

Conducting information-psychological operations by the Russian Federation strives to shape the desired behaviour of the government, various organisations, groups and individuals aimed at splitting the society between regions, between Russian-speaking and Ukrainian-speaking people, between people at the frontline and the people in the safe territories, as well as at dividing the Ukrainian people by their national and political sentiments [10].

In our analysis of the recent events taking place in October and early November 2022, we will distinguish the most relevant topics that have spread across the

Internet media environment and sparked active discussions. For this analysis, we have selected the three most popular information portals according to the rating based on the survey of the OPORA Civil Network (Fig. 3) and the channels which are the largest in terms of involved audience, topics and regularity in updating their content: in Telegram – "Livyi bereg" (audience – 6,821,000), "Telegramna sluzhba novyn" (audience – 1,082,827), "Top News" (audience – 1,067,941), "Ukraine Now" (audience – 903,006), "Realna Ukraina" (audience – 355,302), S. Sternenko's channel (audience – 305,005), "Zakhid 24/7 ONLINE" (audience – 79,575), on YouTube – "Fabryka novyn" (audience – 2,068,000), 2+2 TV channel (audience – 685,000), 24 Channel (audience – 179,008), on Facebook – TSN (audience – 1,303,705), "Ukrainska pravda" (audience – 638,480). Having analysed the media content, we can distinguish the most relevant topics within the analysed period and their influence on the audience:

1. ***Use of nuclear weapons.*** Insider information in social media on possible dates of such use, photo and video materials about the movement of nuclear warheads; discussions about dirty bombs. Let us quote some examples. For instance, on a **Telegram channel,** audience reactions were generated by the following messages: *"NATO Secretary General Jens Stoltenberg – next week the Alliance is going to hold nuclear deterrence exercises", "Russia has sent bombers capable of carrying a nuclear charge to the border with Norway", "Russia is capable of completely destroying the United States and Europe with nuclear missiles in less than 30 minutes and vice versa. A surprising number of people do not know this. Elon Musk"* (Telegramna sluzhba novyn); *"Scholz calls nuclear strike a red line that must not be crossed, and called to refrain from it", "The aggressor will continue conventional military operations to hold the previously occupied territories but is unlikely to use tactical nuclear weapons"* (Top News); *"Inspectors of the International Atomic Energy Agency (IAEA) have completed their in-field verification activities at three locations in Ukraine at the request of the Government of Ukraine. The Agency did not find any indications of undeclared nuclear activities", "...nobody in Ukraine has created, or is currently creating a dirty bomb. The only thing which is dirty in our region now is the heads of the people in Moscow who have unfortunately seized control of the Russian state and are now terrorizing Ukraine and the whole world", "Biden questioned whether Russian leader Vladimir Putin was sincere in saying he has no intention of using nuclear weapons in Ukraine"* (Livyi bereg). On one **YouTube channel**, videos appeared with titles like *"Surovikin will start the nuclear war. Samus: Putin will not even know when the new war starts", "Russia will turn the south of Ukraine into a nuclear desert!"* (Fabryka novyn). On

Facebook: "*A Russian nuclear strike would "almost certainly" trigger a "physical response" – a NATO official*", "*The USA and Europe will not strike back with nuclear weapons: experts announced consequences for Ukraine*", "*The USA and NATO will enter the war if Putin uses nuclear weapons: Fukuyama describes the scenario*", "*The UK defense minister condemned Macron's statement about Russia's use of nuclear weapons*", "*Lukashenko made a cynical statement concerning the war in Ukraine and nuclear strikes: Russia owns the most up-to-date kinds of weapons (TSN).*" Having analysed the content where the issue of nuclear weapons is kept up to date, we can observe the polarity both in the views expressed by experts and in the coverage of this topic on various media platforms: it ranges from being "last step before they are used" and drastic consequences for Europe and the USA to doubts concerning the probability of their use, mentioned locations from where they can be launched, possible NATO response against Russia in case of a nuclear strike, and varied opinions on the issue expressed by politicians across the globe. We can observe a kind of emotional shattering of the audience: from fear, annoyance, aggression to manifestations of critical thinking ("How can you make such assumptions?", "We should all watch less of such experts") and trust in the Armed Forces of Ukraine ("Everything depends on the AFU").

2. ***Belarus***. The topic that is most disturbing to the audience located in the north and west of Ukraine: troop movements, probable invasion dates. On the top-ranking **Telegram channel**, we can find such posts: "*Belarus will not place military bases on its territory – so far*", "*Some mass media report kamikaze drones spotted in Belarus*" (Realna Ukraina), "*Putin demands Lukashenko to join the war against Ukraine*", "*Meanwhile, Belarus continues its mock action, that is, preparations for a probable invasion from Ukraine*" (Zakhid 24/7 ONLINE); "*head of Rivne Regional State Administration Vitaliy Koval claimed there is no threat of invasion from Belarus*", "*If Belarus rushes here, it will be the end of Lukashenko – Oleksandr Karpiuk, Ukrainian marine*", "*The Russian Federation may resume its advance on the western part of the border between Belarus and Ukraine in order to cut the weapons supply lines from the Western partners*" (Livyi bereg*); "Belarus is preparing for invasion from Poland" – Head of Belarussian KGB I. Tertel*", "*Smerch and Tochka-U in Belarus. They are being concentrated not only on the Belarussian borders, but also along the border with Poland*", "*There will be no offensive, but new missile launches: Arestovych on the situation in Belarus*" (Telegramna sluzhba novyn). **On YouTube channels:** "*France threatened Belarus for a possible attack against Ukraine*", "*British intelligence services brings surprising news – Belarussians are pushing to Ukraine along with Russians*", "*To Lutsk – and Lviv! Zhdanov forecasted*

a repeated invasion from Belarus" (24 Channel); *"Is Belarus going to attack? Vehicles with suspicious marking have been spotted", "An attack from Belarus is to Ukraine's advantage: it will be easier to destroy the enemy"* (Fabryka novyn). **On Facebook**: *"Putin wants to move troops to Belarus to strike against Kyiv and Volyn", "Belarus accused Ukraine of 'preparing a strike' The Ministry of Foreign Affairs supposes it to be a part of Russia's provocations"* (TSN), *"Belarussian Ministry of Defense claimed the West is preparing to wage war against them and the Russian Federation"* (Ukrainska pravda). Some of the above-mentioned headlines resort to frightening the readers, tend to exaggerate and practice excessive emotional colouring, and there is no confirmation in the text of the news item itself. They inform the audience about troops being trained and prepared in Belarus without giving specific dates of the attack on Ukraine and contain threats as to the plans to strike Poland. What is more, considerable support from Western partners, trust in evident defeat of invaders from Belarus and the successful counteraction of the AFU evoke confidence and optimism in Ukrainians. In comments, one can observe sarcasm, irritation, indignation, threats – and at the same time, support for the Armed Forces of Ukraine and trust in the Ukrainian army.

3. **An increase in terror** can be observed in the information that covers destruction, torture, pillaging etc. **The Telegram channel** of "Ukraine Now" publication posted content concerned with the cruel treatment and murder of the civil population by Russians: *"Russians killed 428 children in Ukraine", "Wherever the invaders come, they only leave pain, destruction and losses behind. In the village of Vysokopillya in Kherson Region, there used to be 4,000 people before the war, but only 2,000 now", "Russian warships caused the death of 50,000 dolphins in the Black Sea"*. "Telegramna sluzhba novyn" reports: *"Russia is losing on the battlefield and switches to terrorizing the civil population by destroying critical infrastructure", "Russian sadists arranged torture chambers in the basements of the station and a shop", "Torture, pillaging, rape, followed by occupation of the Zaporizhzhia Nuclear Power Station – these actions suggest that we cannot exclude the risk of the use of nuclear weapons" (President Zelenskyi).* **On YouTube channels:** *"Tabakh: Russia has found a killing machine for the war – they were trained and then betrayed by the USA", "Russia wants a cold and bloody winter in Ukraine", "There will be more missiles"* (Fabryka novyn); *"Bratchuk: The Kremlin is blocking exports (suspension of the grain deal), "Zelenskyi warned about the catastrophe. Russia has undermined the Kakhovka Hydroelectric Station"* (24 Channel). **On Facebook**: *"Bombing peaceful cities is the act of terror to exert pressure on Europe and Ukraine" (President Zelenskyi), The house blown up and the woman kidnapped: in Kherson Region,*

invaders disposed of the couple who supported Ukraine" (TSN); *"Stopping the Zaporizhzhia Nuclear Power Station brings a radioactive catastrophe closer", "Invaders ordered Kherson residents to leave the city and cross to the left-bank of the Dnipro River"* (Ukrainska pravda). Comments from the audience reflect pain, trauma, hopelessness, despair, exhaustion, fury. Such media content devastates, leads to absolute helplessness and powerlessness, as well as repeated traumatisation of the people who lived or still live in the territories affected by Russian acts of terrorism. The audience express a convincing idea that Russia is a terrorist state that has brought death and destruction to Ukraine. Millions of people live in the state of permanent stress and anxiety, but it should be understood that breaking the Ukrainians' inner strength is actually the true goal of Russia. That is why the informational resilience of the audience is most vital, because despite the crisis situation in our country, we need to withstand the enemy's information terrorism (i.e. manipulation of people's consciousness) that is implemented by placing the people into a controlled information environment.

4. ***The energy problem in Ukraine is*** acute, but we should consider the fact that the enemy's main tactical aim is to attract attention to a problem and create a lot of information noise, and the strategic target is to frighten and demoralise people for the sake of securing certain results. This is how this topic is covered in the information environment: on **Telegram channels**: *"The President's office made public their plan to counter "the energy genocide" on the part of Russia", "Residents of Kyiv Region need to prepare for emergency power cuts for an unknown period of time", "If voluntary reduction of energy consumption is insufficient, power cuts will have to be forced in Ukraine"* (Livyi bereg); *"We ask the Ukrainian refugees not to return home until the spring. Unfortunately, they won't bear it. If you have an opportunity, stay abroad for the winter" (ex-Minister I.* Vereshchuk), *"Evacuation of the people in case of a critical situation with energy supply is an emergency action that the government may take only after all the others – D. Shmygal"* (Telegramna sluzhba novyn). **On YouTube channels,** one can find video content with such headlines as *"Putin's plan with strikes against thermal power plants is revealed! Who will be deprived of electricity – Sakharuk from DTEK" (Fabryka novyn); "How long will Ukrainians have to stay in the darkness? How can you help in the days of the power crisis?", "Power cuts across the country due to Russian attacks"* (24 Channel). **On Facebook:** *"Ukrainians started to save electricity voluntarily", "Power cuts in Ukraine are divided into two types, stabilization and emergency"* (TSN); *"The capital's mayor Vitaliy Klychko doesn't exclude a complete blackout scenario in Kyiv with the absence of electricity, heating, water and communications, and*

> *calls for Kyiv residents to create emergency supplies and consider temporarily moving to the country", "After massive Russian strikes against thermal power plants and combined heat power plants started on 10 October, the situation has changed completely. Electricity has become arguably the most sought-for product these days"* (Ukrainska pravda). Fear, panic, anxiety, despair – the emotional state of the audience that can be observed from comments made in the form of text using signs and characters. Such reaction from the readers is easy to understand as the news concerns critical infrastructure that ensures satisfaction of people's basic needs. However, in their comments people try to solve this problem in a rational way and prepare themselves to the possible difficulties. It leads to expression of critical thinking, creative inventions, and conscious energy-saving tips.

Having analysed media content on information platforms most popular with the audience, we can conclude that content users become subject to emotional swings as they systematically view news with absolutely opposing opinions and forecasts, with positive facts and terrible events posted alongside each other. This results in such emotions as negative attitude and stress, which are stronger and more important for survival. Because of such conditions, the audience may become voluntary boosters of the information preferred by the enemy, as they spread such information in their circles of communication, groups and collectives at work. The enemy attempts to neutralise, in all kinds of ways, the energy of war ("we will withstand / win / defend") that awoke in the mass conscience, and establish reflexive control – that is, to influence the emotional and cognitive aspects of the audience with various information and manipulative technologies in order to change the Ukrainians' behaviour strategy in the way that would benefit the aggressor.

It should be noted that the emotional component of informational resilience is the most important one, since stabilisation of one's psychological situation after encountering negative information is the road to developing internal motivation, inner power which helps the audience to cope with all negative sides of life.

MINDY Foundation for adaptation to peaceful life and Kantar Ukraine, together with the Institute of Social and Political Psychology of the National Academy of Educational Sciences of Ukraine have conducted a survey across all Ukraine to explore psychological state of the people in conditions of the large-scale war. According to the survey results, they have determined the level of psychological well-being of Ukrainians (Fig. 4). In spite of the difficult circumstances of the large-scale war, the subjective evaluation of the psychological well-being of Ukrainians is 6.7 points out of the maximum value possible, 9 points. This is typical for all demographic groups and regions. The largest contribution to the feeling of psychological well-being consists in self-esteem (the respondents' positive attitude towards themselves). Meanwhile, instrumental elements of well-being, such as ability of emotional self-regulation and overcoming complicated emotional states (emotional component) or the ability to cope with challenges of life (persistence component) are rated at lower levels. This situation suggests that the stage of psychological mobilisation caused by the war is still active in the society, but it may lack resources to maintain this state for longer periods [7].

Fig. 4. The level of psychological well-being of Ukrainians under wartime conditions

It is a well-known fact that people in stressful situations are especially easy to manipulate during information warfare. When exhausted, it is easiest for human brain to accept all signals that forebode despair and hopelessness, which is extremely dangerous during information-psychological attacks.

For the purpose of exploring the emotional background and responsible media consumption under wartime conditions, a survey (online questionnaire survey) was conducted on the audience of the Ternopil Region aged 17 to 70 in October–November 2022. The survey involved a total of 187 respondents.

When answering the question of what channel they receive information from, the majority of respondents – 90.4 % – choose Telegram. The second most popular source is Instagram (50.8 %), while relatives, friends and colleagues are the news source for 43.3 % of the respondents, television is favoured by 36.6 % of people, and social media preferences spread from TikTok – 34.2 % to YouTube – 29.4 % to Facebook – 21.9 % and Viber – 16.6 %, while radio (6.4 %) and Twitter (4.8 %) appeared at the bottom of the rating. The least percentage of the audience trusts printed media – 3.2 % (Fig. 5).

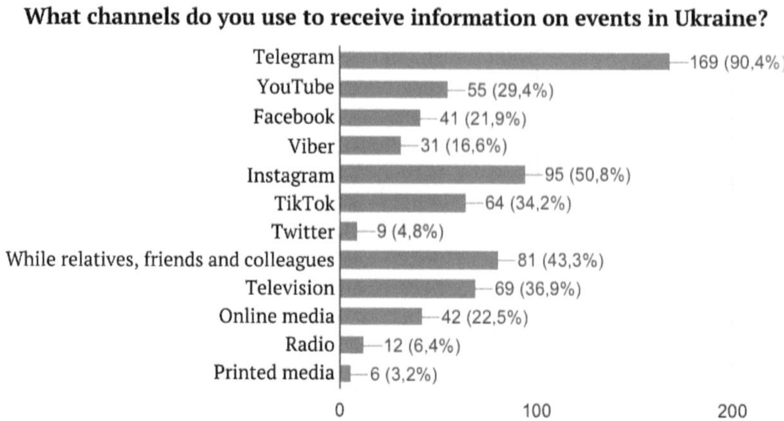

Fig. 5. The channels used by the audience to receive information on events in Ukraine

Hence, social media take the leading positions as the source of news. This is related to their promptness in presentation of news and the declining trust to all conventional national media.

The key advantages of Telegram chosen by 90.4 % of the respondents under wartime conditions is the chronological nature of its feed, easy-to-use interface, active transition of government agencies (including security-related authorities – e.g. air raid alarm notifications) to this medium, availability of multifunctional chatbots, swiftness of communication, actual absence of content moderation etc. Among the disadvantages on which the audience should focus on is the absence of information about channel administrators, except the data which they have provided publicly, and there is also no opportunity to track the methods of promoting content. That is why the audience should use several channels to get accurate information.

To access news, 66.3 % of respondents choose 2–4 sources, while 5–6 sources are chosen by 21.4 %, 7 or more sources by 11.2 % and one source by 3.2 % of respondents (Fig. 6). To protect from information noise, we recommend selecting the most trusted channels and reducing information intake to these. However, a single source with an untarnished reputation will not suffice. It is believed that 5–6 main sources of information about the current situation in the world and in Ukraine is the optimal selection.

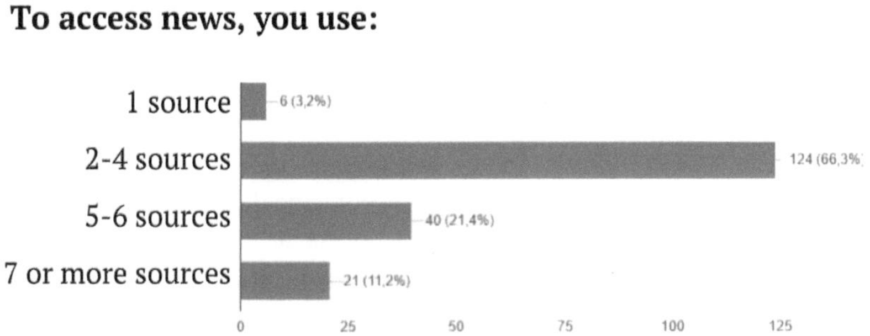

Fig. 6. The number of sources used by the audience to access news

Since a considerable increase in the volume of fake news, disinformation and manipulations can be observed in Ukraine's information environment, it is important that the audience should only trust official sources. On the whole, 58.8 % of the respondents receive information from official sources of specialised government agencies, whereas these sources are used sometimes by 33.2 % of the respondents, while 6.4 % of the respondents do not use them at all. Not all respondents (5.3 %) know which official sources can be trusted or on which platforms these can be found, so it was difficult for them to choose their answers (Fig. 7).

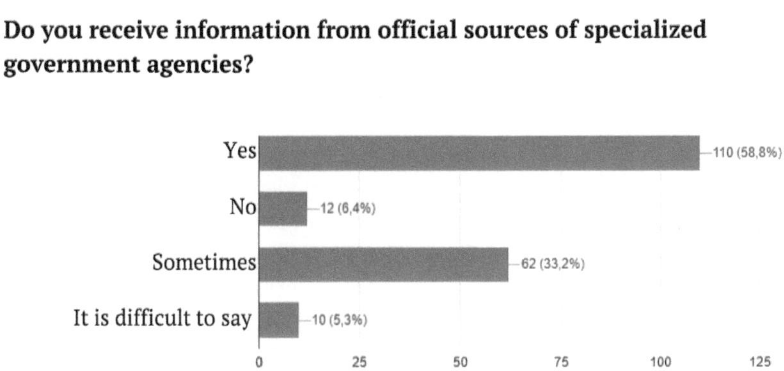

Fig. 7. Does the audience receive information from official sources?

Under wartime conditions, especially in extreme situations, much false information is spread, especially when there is a lack of facts concerning the events taking place in the society. People seek information and try to receive it from any sources, and hence may encounter fake news and spread it.

When asked "Do you always believe published information?" only 11.2 % of respondents were positive, whereas 63.1 % were negative, while 26.7 % were doubtful as to its faithfulness (Fig. 8). The present-day challenge is to develop media literacy, which is based on a model encouraging the audience to reflect on what they watch, listen and read. The task of media literacy consists in transformation of media consumption into an active and critical process that apparently influences formation of informational resilience.

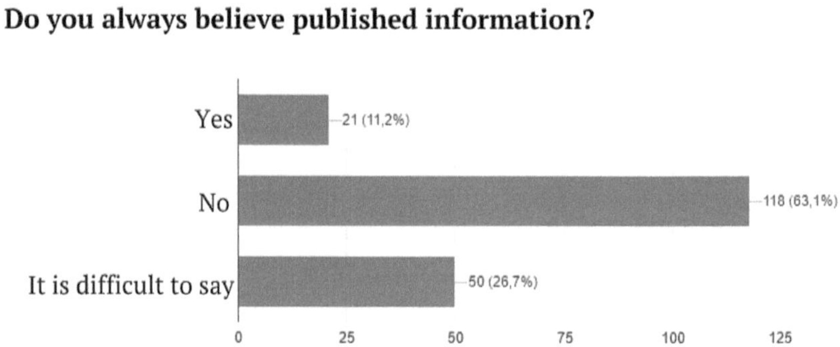

Fig. 8. Trust in information

In this most difficult period of Ukraine's history, the audience experiences changed in mood and behaviour during the events caused by military operations. According to the survey, the majority of respondents experiences negative emotions: 47.6 %, including 47.1 % irritation, 33.2 % fear and 24.1 % despair. Despite the country's critical situation, a part of the respondents is positive, in particular, optimism and trust in victory are typical for 24 % of the respondents. Among the audience there are people in the state of anticipation (37.4 %) or acceptance of the situation 38 %, people interested in how the events may develop 15 %; those who are surprised 8.6 %, and there was a single (1) respondent who felt indifferent (Fig. 9). Emotionally coloured information affects the recipient's condition that may change drastically in such directions: satisfaction, dissatisfaction, dual (ambivalent) or uncertain attitude to reality. Meanwhile, a person can be in a very intense condition, agitated condition or in a calm condition.

Aggression Irritation Anticipation Interest Optimism Acceptance Surprise Fear Despair Other

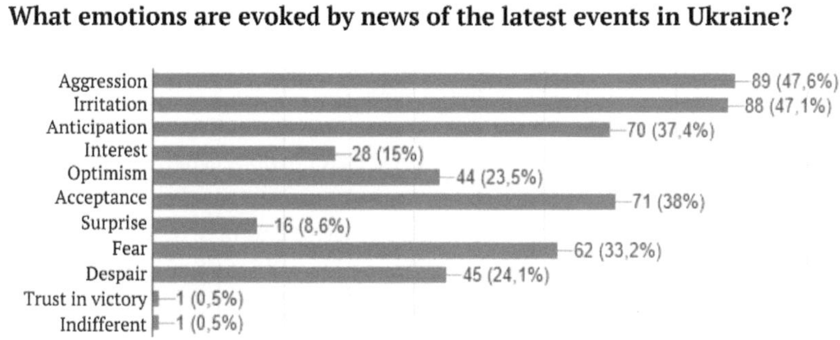

Fig. 9. The emotional field of Ukrainians in wartime conditions

The reaction of the audience to media content and responsibility for spreading it is one of the elements of informational resilience. When asked "Does a post cause an immediate desire to react in any way (share it, comment on it in social media etc)?" 54 % of the respondents answered "sometimes", 35.3 % "never", whereas 15 % of the respondents admit to promptly reacting to posts (Fig. 10). Hence, a major share of respondents tends to promptly share media content and react to it without applying a critical approach to analyse the information and verify it. This makes identifying information operations more complicated, creates tensions and evokes manifestations of hate and dissatisfaction in social media etc.

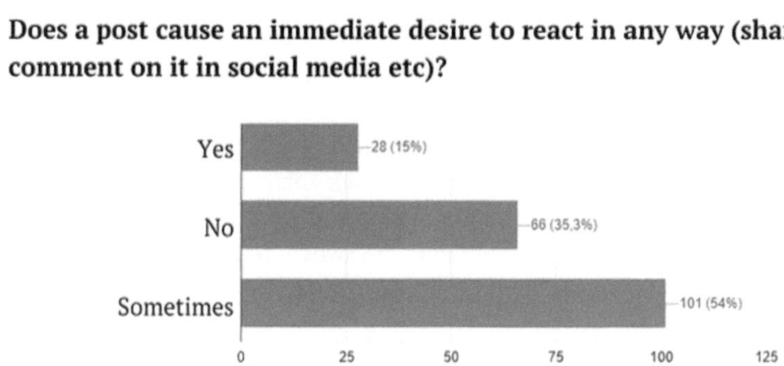

Fig. 10. Audience reactions to posts

Being informed is important, but it is also responsible not to harm the others. When asked "How do you react to emotionally colored information?" 48.7 % of the respondents demonstrate emotional stability and take a critical approach to analyse information about the current situation in the country. Meanwhile, 38 % of the respondents share it with relatives and close people, whereas 11.2 % want to inform as many people as possible about the current events (Fig. 11). It should be noted that in a typical situation it is manipulative and fake news that tends to "hook" the audience by using emotional levers. In such situations, informational resilience consists in manifestation of emotional intellect and critical thinking.

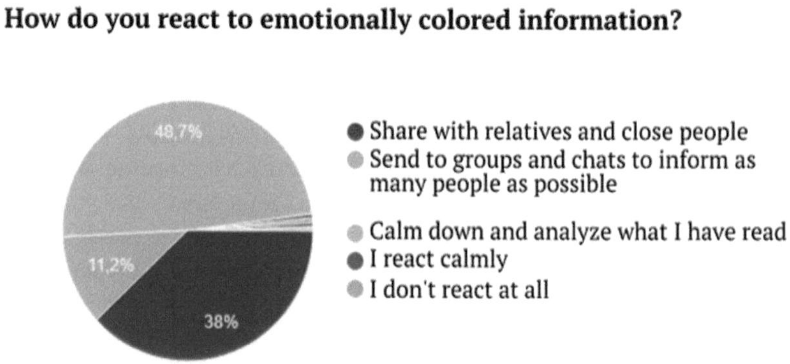

Fig. 11. Audience reaction to emotionally coloured information

Respondents use a variety of methods to stabilise their emotional state. In particular, they focus on the greatness and power of the Ukrainian people bound to win (51.3 %); share their feelings with relatives and close people (44.9 %); switch to work (33.2 %); pray (27.3 %); write about their feelings on social media (7 %); read books, write poetry, listen to music (1.5 %); analyse the situation and forecast the subsequent events (1 person); meditate or limit the amount of news to reduce the flow of negative information (1 %); switch to games, walks, or memes (1.5 %); or do not use any methods to improve their state (1.5 %) (Fig. 12).

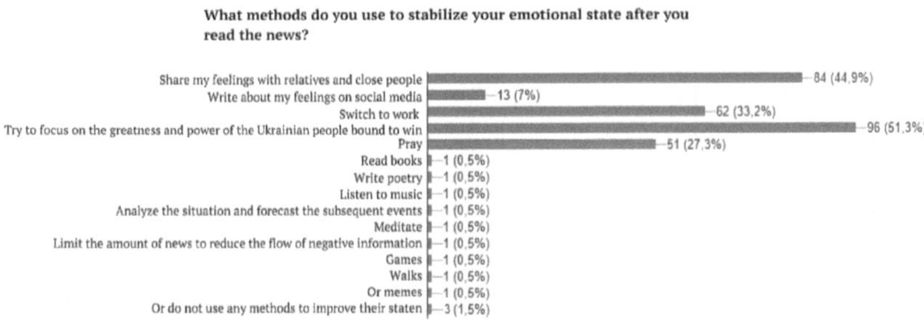

Fig. 12. Methods used to stabilise emotional state

None of these methods can be evaluated as "correct" or "incorrect". For each person, there is a specific method that appears to be most effective for stabilising their emotional state. However, it is worth noting that reading a piece of news evoking many emotions needs to be postponed for some time.

We would like to add that media content exerts an emotional influence on the audience and acts as an external irritant, later becoming an internal irritant when it arises from emotional memory, and it provides sufficient grounds to consider the emergence of internal triggers as a consequence of mass media influence. Later on, negatively coloured information covering urgent issues of the day tends to affect recipients' psyche by evoking anxiety, fear, panic and aggression (see Ref. [4], p. 36). That is why it is worth focusing attention on the main factors that facilitate stabilisation of one's psychological state during the war, while considering the emotional component as the leading element of informational resilience. These factors include:

- transformation of emotions. For example, from pain and fear caused by the destruction of the war to pride in the military and the steadfastness of the Ukrainian people;
- focusing on facts. Aid from Western partners facilitates our victory. This will not bring the dead back to life, but the responsibility for the committed crimes is imminent;
- transformation of negative emotions into actions. This category includes volunteering, posts in social media, and prayers.

The rapid rate of spreading news and the belief that it requires prompt reaction, encourages "quick thinking" (reflexive and emotional) as opposed to "slow thinking" (rational and sensible). Under wartime conditions, critical thinking skills are extremely necessary since they prevent people from getting lost in the environment oversaturated with information, and from falling for manipulative tricks. Critical thinking is the skill of asking questions, substantiating one's opinion, evaluating and interpreting information, viewing events in parallel with present-day realities, and analysing the preconditions and causes of events. This is also the ability to apply rational methods to search, select, systematise and verify text, photo, and video materials, use special online services that can help identify manipulations in content (see Ref. [4], p. 168).

To form a responsible attitude in the audience to media content and improve their resilience to the destabilising effects of fake news, it is necessary to distinguish features of false information and manipulation of facts, which include the following:

- excessive emotionality;
- exaggeration;
- partial concealment of information;
- substitution of notions;
- overrating or underrating;
- incorrect comparison of data;
- presence of quotes without context;
- annoyingly persistent positive-negative background;
- references to the past only;
- references to Wikipedia information only;
- avoidance of discussion concerning difficult topics;
- viewpoint and hidden intentions of the author;
- correct statistical data accompanied by misinterpretation;
- reputation of the news source;
- affiliation of the news source with specific people and their political background.

As noted by Shevchenko, the necessary means of verifying information accuracy are as follows: comparison of information from several sources, distinguishing between facts and value judgements, identifying the primary source (for verbal and non-verbal content), checking content and context as well as statistical data by open official sources, exploring the circumstances of an event, comparing information in various sources by known facts related to this topic, searching for the primary source of quotes and additional information about an event, expert evaluation, checking photos according to such criteria as date, weather, landscape, people, clothes, shadows, as well as scrupulous treatment of every detail pertaining to the topic (see Ref. [11], pp. 150–151).

In order to hold the information frontline, to be informationally resilient and preserve one's resources during emotional instability which is manifested under wartime conditions, Otar Dovzhenko, an editor, journalist, media coach and media explorer, advises using the following opposition mechanisms [1]:

1. Do not trust unconditionally. One should trust only quality and responsible media. However, we should remember that any piece of news during the war may be inaccurate, or never confirmed.
2. Be careful with interpretations. Those who generate fakes and propaganda materials know that news accompanied by photos, videos or audios are more likely to be trusted. Meanwhile, comments or introductions to such photos, videos or audios often provide a ready-made interpretation therefore programming the recipient to accept the materials in a certain way.
3. Do not trust any retellings. Research conducted by the Detektor Media Public Organization and the Democratic Initiatives Foundation has demonstrated that receiving information by word of mouth is one of the most widespread ways of learning news. For instance, 28 % of the respondents learn news primarily from friends and acquaintances. Important news received from other people needs to be checked in official sources.
4. Do not consume information from hoax news websites. Toxic materials make people sad, amplify their anxiety, make them lose the feeling of reality, and provoke arguments where both sides struggle to convince the opponent.
5. Monitor your reactions. A destabilised, fear-paralysed and confused person is of doubtful value for their homeland.
6. Do not share unverified information, since fakes can go viral and cover the audience of millions.

Conclusion

The war in Ukraine is the field for information and psychological confrontation where an important role is played by informational resilience of the audience and their responsible media consumption. These days, Ukrainians live within an information bubble which leads them to information overloading characterised by features such as: fatigue, decreasing functions of self-control and ability to consider all facts, and as a result, the choice is made subconsciously rather than consciously; promptness, which reduces the accuracy in processing information; ignoring certain kinds of information while other data is processed; unconscious decision-making resulting from exhaustion of resources; avoidance and inactivity, evasion from solving tasks related to information analysis. Because of the above, the Russian Federation implements information-psychological operations to influence public sentiments and change the vector of actions taken by society. Having analysed the recent events (October – early November 2022), we have distinguished the most relevant topics actively discussed in the media environment. They are related to the use of nuclear weapons, invasion from Belarus, the increase of terrorist attacks, and the energy problems in Ukraine. Having analysed the media content, one may conclude that the audience is now affected by emotional swings manifested as changes in mood, feeling of constant anxiety and discordant emotions. That is why it is expedient to consider the emotional component as a leading element of informational resilience and outline the main factors for stabilising one's psychological state during the war. Having conducted a survey of the audience in Ternopil Region for identification of their emotional background and responsible media consumption in wartime conditions, we can state that the recipients mostly experience negative emotions (aggression, irritation, despair), but they try to stabilise their emotional state by using various methods, dominated by thoughts of greatness and power of the Ukrainian people bound to win, and by communication with their close people and relatives. Responsible media consumption is evidenced by the fact that most respondents receive information from channels provided by official authorities and use two to four sources to access news. To win the cognitive war, one needs critical thinking skills, which can help to distinguish the signs of false information and manipulation of facts, and master the means required to verify the accuracy of information. With mechanisms for combating disinformation, informational resilience is manifested, which is an important factor for securing victory not only on the information front, but in the actual war as well.

References

[1] Dovzhenko O. Yak chytaty novyny pid chas viiny. MediaLab. https://medialab.online/news/novyny-pid-chas-vijny/.

[2] Dutsyk D. "Yak transformuiutsia ukrainski media u vohni viiny". Journalism teachers' academy. https://cutt.ly/gNePkIL.

[3] Koval Z. "Informatsiino-psykholohichna stiikist osobystosti ta sotsiumu v literaturi: analiz doslidzhen ta publikatsii". Teoriia ta istoriia derzhavnoho upravlinnia, 2019. Tom 2. № 78. p. 18. https://doi.org/10.35432/1993-8330appa2782019178957.

[4] Kuznetsova O. "Psykholohichni tryhery v ZMI: poniattia, osoblyvosti, vydy, protystoiannia yikh vplyvu". Visnyk Natsionalnoho universytetu "Lvivska politekhnika". Seriia: Zhurnalistski nauky. Lviv: Vydavnytstvo Lvivskoi politekhniky, 2017. № 883. pp. 33–38.

[5] Mediaspozhyvannia ukraintsiv v umovakh povnomasshtabnoi viiny. Opytuvannia OPORY. https://www.oporaua.org/report/polit_ad/24068-mediaspozhivannia-ukrayintsiv-v-umovakh-povnomasshtabnoyi-viini-opituvannia-opori.

[6] Melnyk I. "Informatsiina stiikist Ukrainy: vybir trendiv masovoi kultury ta yikh vplyv na suspilnu svidomist I stratehii derzhavnoho upravlinnia". Vcheni zapysky TNU imeni V.I. Vernadskoho. Seriia: Publichne upravlinnia ta administruvannia, 2021. 32 (71). № 2. p. 69.

[7] Pres-reliz – Doslidzhennia psykholohichnoho stanu naselennia v umovakh povnomasshtabnoi viiny. Instytut sotsialnoi ta politychnoi psykholohii Natsionalnoi akademii pedahohichnykh nauk Ukrainy. https://ispp.org.ua/.

[8] Romanchuk O. Psykholohichna stiikist v umovakh viiny: indyvidualnyi ta natsionalnyi vymir. Ukrainskyi instytut kohnityvno-povedinkovoi terapii. https://i-cbt.org.ua/resilience_ukraine/.

[9] Slovnyk z mediahramotnosti. Natsionalnyi proiekt z mediahramotnosti «Filtr». Ministerstvo kultury i informatsiinoi polityky Ukrainy. URL: https://filter.mkip.gov.ua/wp-content/uploads/2022/10/slovnyk.pdf.

[10] Shevchenko V. "Faktchekinh i veryfikatsiia u zhurnalistskii roboti". Obraz, 2018. Vyp. 1 (27). pp. 140–153. http://doi.org/10.21272/2415-8496-2018-1(27)-140-153.

Olesia Medynska

Disinformation Narratives of Social Media in the Context of Social Transformations

Introduction

Disinformation has become a part of the information environment, especially in periods of social transformation. For Ukraine, disinformation remains to be a serious threat as it is one of the main tools applied by the Russian Federation during the war. Disinformation encourages the destabilisation of society and provides ground for conflict-generating situations and potential combat actions. In conditions of a psychologically unstable society, people will accept as true any content which is generated by social media and which corresponds to their emotions rather than to rational thinking.

The analysis of processes taking place in the modern information environment demonstrates an increasing influence of social media on mass consciousness. In conditions of social transformations and states of emergency – revolutions, pandemics and wars – the problem becomes more exacerbated since it is the physical rather than informational security of the citizens that depends on whether the facts they have received are true or false. In this context, identification of disinformation narratives in the social media discourse in order to protect society from their destructive influences and to find the ways of countering them appears to be especially relevant.

Results of the Research

The purpose of this research is to identify the content-related priorities of disinformation narratives in social media in periods of social transformation.

The goal of the research is to outline the definitions for such terms as "narrative", "disinformation", "disinformation narrative"; to present various approaches to classification of disinformation in scientific and practical discourse; to perform monitoring of social media in order to identify disinformation constructs; to classify disinformation narratives; and to suggest ways of countering them.

To achieve the goals we have set, the systemic-functional analysis was used, which envisaged the application of a series of methods and approaches. General research methods: analysis of academic literature was used to outline the conceptual structure of the research; descriptive method – to observe and interpret disinformation narratives; classification method – to distinguish various kinds of disinformation narratives; content analysis method – to explore the disinformation content in social media; synthesis and generalisation – to systematise and separate results of analysis of disinformation narratives in social media.

Disinformation and its discursive characteristics have been a subject of research for both Ukrainian and international scientists, media experts, and information environment explorers. The problems of conceptualisation of the term "disinformation" are an integral part of studies undertaken by Floridi and Fallis. The role of disinformation in information-psychological operations was explored by Petryk, Shtokvysh, Pocheptsov and Prysiazhniuk. Methods of countering disinformation were considered in the works by Stray, Giussani, Fursai, and Danylenko. The issue of studying the role of narratives in the consciousness of modern society was the focus of research works by Hutsol, Isakova, Karpyleko, Surina, Sushyi, and Shkil. The interconnection of mythological and ideologemic constructs in the hybrid warfare was studied by Zelenin, Kononenko, Magda, Mylchenko, Nagorna, Perekhnoskyi, Pocheptsov, Rozumnyi, Stiazhkina, and Yavorska. Influences on mass consciousness through the concept of information warfare were explored by Chernova, Grushyn, Lysenko, Diligenskyi, Olshanskyi, Fromm, Panaryn, Libicki, McLuhan, Kara-Murza, Moscovici, Linebarger, Senchenko, and Varyvoda. At the same time, in spite of the incontestable value of their contributions, the scientific discourse lacks research works focusing on disinformation narratives of social media, which suggests the choice of this topic.

The point of departure of this research is clarifying the essence of the notion of "disinformation" since without first doing so, one cannot identify disinformation narratives and understand the nature of their creation. Let us study the definitions of this notion, "disinformation", as given by different researchers, and its theoretical construct. Fallis, an American scientist, remarks that disinformation is non-accidentally misleading information [6]. The Oxford English Dictionary interprets disinformation as "false information that is given deliberately" [9]. The European Commission defines disinformation as "false or misleading content that is created, presented and spread with an intention to deceive or secure economic or political gain, and which may cause public harm" [5]. A Ukrainian researcher, Pocheptsov notes that disinformation features "a special emotional breadth that facilitates spreading disinformation narratives across large masses of people"; it is difficult to disprove; it promotes an "imagined virtual object which engages actively in breaking the conventional worldview"; it encourages the object of disinformation to commit certain actions which put the subject of disinformation to advantage; a "deliberate deception" [10]. The educational platform EdEra in cooperation with IREX provides the following definition: disinformation is "manipulative audio / video content created deliberately. It combines two features: being false and having the intention to cause harm" [3]. Media experts from the Institute of Mass Information understand disinformation as "purposeful spreading of false information, use of manipulations for deception and misleading, and provoking loss of trust to any source of information" [8], while distinguishing such qualities of this phenomenon as deliberate dissemination of false information and teleological nature. Researchers from the online education platform Prometheus emphasise that it is "false, manipulative and/or deceptive information which is spread deliberately to achieve a certain political goal" [13]. Analysts from TEXTY.ORG.UA believe that the notion of "disinformation" suggests an emphasis on a certain pattern of worldview, an opinion – the so-called spin [14]. Experts from Detektor Media note that disinformation is dissemination of false information, distorting facts with the purpose of influencing opinions and behaviour of an individual or society, and making social groups believe in a worldview which is different from reality [4]. Urška Umek, Head of the Department of Information Society, Council of Europe, claims that it is "messages that deliberately manipulate facts and reality with the aim of inflicting damage or gaining financial profit; fabricated or deliberately modified audio/visual content, or specially created conspiracy theories or rumors" [1]. According to Koretska, disinformation is public dissemination of false information which was not checked or was proved to be false, and which has, or may have negative consequences for implementation of people's constitutional rights and/or threaten national security [7].

Having analysed the above-mentioned definitions for the term "disinformation", we can distinguish its conceptual features: significance of information, falsity or deceptiveness, specifically formed data, purposeful spreading of data, the intention to deceive or harm, teleological nature, systematicity, and emotional breadth.

Having explored various scientific approaches to defining this term, one can conclude that disinformation should be viewed both in a narrow and enlarged sense of the word. In the narrow sense it is false information, deliberately prepared and spread in order to achieve a predetermined political, economic or ideological goal. In the enlarged sense, it is a purposeful process of creating, modifying and disseminating information, both false and true, with the purpose of deceiving a person (a group of people) to achieve a political, economic or ideological goal; as a result, human rights / civil rights / rights of the public / rights of state are infringed or may be infringed.

A structural element of disinformation is narrative. Narrative is a "story, interpretation of events that strives to create a desired attitude of the target audience towards the object of such story" [2]; "a plot-and-narration form of organizing the text, a kind of discourse, the subject and meaning contents of which has the status of an event and is characterised by telling a certain story" [15]. The main characteristics of the emergence of narrative is formation of a mysterious plot that consists of selection and arrangement of events and actions which are being told about, and it turns the storyline into an accomplished and complete story that has the beginning, the main part, and the end. The plot is an aggregate of connections, with the help of which events are transformed into a story, or a story emerges from a chain of events, respectively. In other words, a plot appears as an intermediary between an event and a story. A narrative includes all features of a narration text (author, characters, events, storyline etc.) and forms a kind of frame into which real events are inserted in order to be understood and comprehended.

A narrative includes three basic dimensions: semantic (story), mental (representation of the story's world) and discursive (form and style of communication). With the help of narratives, which are actually certain models of stories about true or imagined reality, mass consciousness is manipulated, notions are substituted, and society is integrated into the "pre-determined text" of the narration, while social tensions are created that can transform into certain actions.

With the purpose of accurate separation of disinformation narratives in social media, the following original definition of this term may be proposed. A "disinformation narrative" is a story, a distorted interpretation of reality / event which accumulates and transmits an author's (subjective) systems of senses, and which is aimed at the mass audience with the purpose of influencing their opinions, behaviour and perception.

In the period of social transformations, in addition to strategic (state-generated) narratives highlighting changes in the social reality, the information environment becomes filled with disinformation narratives as well. Within the context of various social and political processes in Ukraine (the Revolution of Dignity, the annexation of Crimea, the COVID pandemics, the economic crisis, the Russian-Ukrainian war, migration processes), it is narratives combining means of ideological, mythological and symbolic nature that form a distorted imagined reality, within the framework of which tailored models of identity, communication, thoughts, judgements and choice of values are offered. Taking into account the transformations taking place in Ukraine, the paradigmatic changes in the principles of forming and implementing state policies, the increasing influence of communities and civil society on political decision-making, using crises and discrepancies in worldview and values of various parts of the society, disinformation narratives are deepening the current destruction and divisions.

A great amount of disinformation is spread through social media where technologies of parallel and cross-cultural communicative strategies are actively used along with other tactics.

Today, social media are quickly gaining popularity while covering immense audiences, which is demonstrated by the results of monitoring surveys performed by Kyiv International Institute of Sociology (KIIS) in 2020 and Kantar CMeter research group in 2022. The survey by Kyiv International Institute of Sociology indicates that respondents distinguish four platforms of social media as their sources of information (see Fig. 1). The dominant social media are YouTube, Facebook and Instagram. Less popular with the respondents are TikTok, Twitter etc. [11].

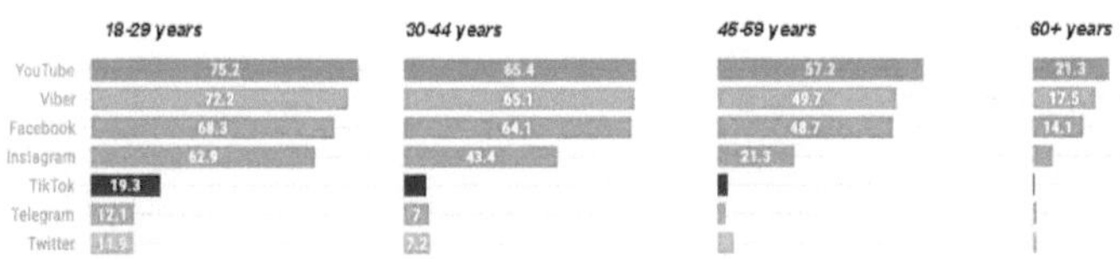

Fig. 1. The level of trust in social media

According to the survey by Kantar CMeter (January–March 2022), the respondents most often chose their sources of information to be such social media as YouTube and Facebook (See Fig. 2) [12].

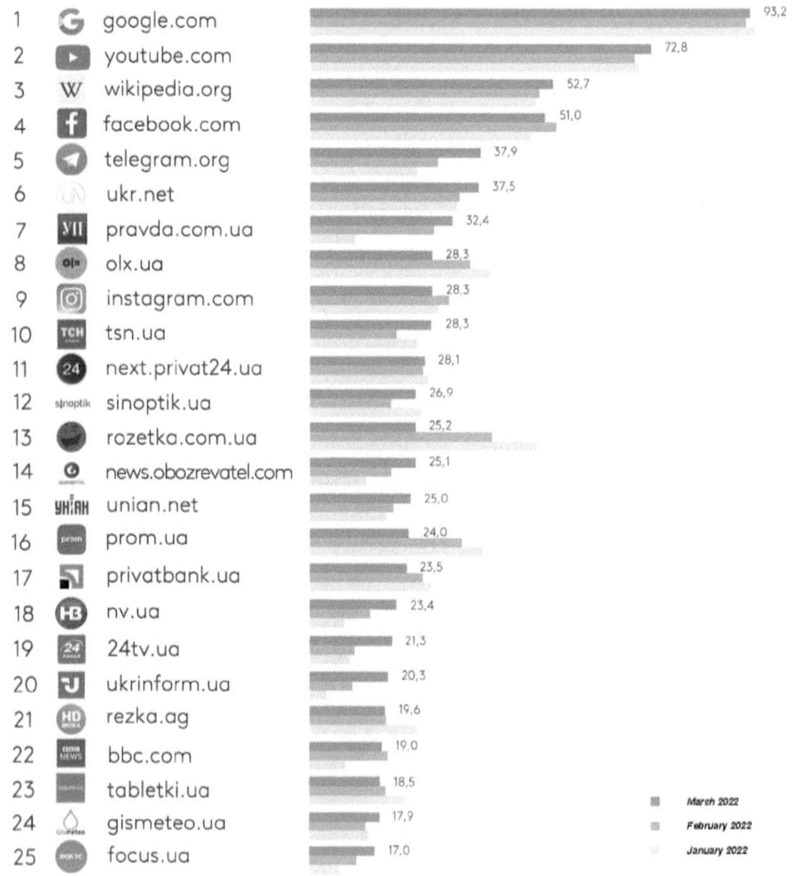

Fig. 2. The rating of popular websites

The key broadcasters of disinformation narratives appear to be the most popular social media – Facebook, YouTube, Instagram, and TikTok.

In the course of our research, we have discovered domination of pro-Russian / Kremlin and anti-Western narratives, whose disinformation flows were intensively spread by the aforementioned social media, with a considerable effect on the people's mass consciousness.

The objects of disinformation include the country's government, the Armed Forces of Ukraine, volunteer battalions, the country's population, national minorities, social groups, international organisations (NATO, UN, EU, WEU, IMF, CE, OSCE, IAEA), governments and population of the world's leading countries.

Let us distinguish the main features of disinformation narratives in social media:

1. Purposefulness.
2. Being based on some "falseness" which can be checked.
3. Being integrated into certain worldview patterns, and conformity to certain beliefs.
4. Construction of a specific hyperreality where people and physical objects are incorporated.
5. Being related to socially significant events, including socially important information which is sensitive or of symbolic value.
6. Irrationality, emotionality, evaluative nature, expressiveness and persuasiveness.
7. Negative (socially dangerous) consequences (anti-vaccination or imperialist motives).

The following types of disinformation can be distinguished in social media:

(1) deceiving a specific person or a group of people by providing misleading information as the grounds for making a particular decision;
(2) modification of the information flow which consists in selective presentation of information;
(3) white noise – the technique of framing true information with false versions which are supported by certain evidence, facts and witnesses.

The disinformation narratives analysed here can be classified as follows:

1. *By the motivation to create them*: political, financial (subjects construct disinformation materials for financial gain) and socio-psychological motivations (purposeful creation of disinformation to achieve informational-psychological influence which forms a psycho-emotional response from the object of such influence, and it results in changes in perception of reality, correction of an individual's worldview in the direction which is advantageous for the subject of disinformation).
2. *By the goals of informational-psychological influence:* gaining an information advantage, changing people's worldview, disorganisation and disorientation, elimination of Ukraine's information structure, destruction of Ukrainian values, and protection of pro-Russian views.

3. *By territorial principle:* local (concerning a certain population centre, city, district or oblast), regional (concerning a certain region within one country), national, international (concerning several countries).
4. *By topic*: political, social, cultural, religious, historical.

Political Disinformation Narratives

Having monitored social media, one can state that in the circumstances of social changes, particularly in the periods of the Revolution of Dignity, annexation of Crimea, ATO/JFO, the COVID-19 pandemics, the full-scale Russian-Ukrainian war, migration processes the most frequently broadcasted disinformation narratives were of political topics.

Disinformation narratives are characterised here as follows: defining objects of disinformation, separating metanarratives, formulating their goals, identifying derived sub-narratives and channels of dissemination.

I. *Disinformation object:* the Ukrainian state and its institutions.

Metanarrative: Ukraine is a dysfunctional state / Ukraine is a failed state.

The goal of the metanarrative: damaging Ukraine's image and authority.

Examples of derived sub-narratives: "Ukraine is a failure of the state"; "Ukraine is the land of chaos, radicals and Nazis"; "Ukraine is a mess"; "Ukraine is dominated by unprecedented neo-Nazism and antisemitism"; "Ukraine has no freedom of speech"; "In Ukraine, Russophobia dominates, and Fascism flourishes"; "Ukraine causes a global food crisis."

Dissemination channels. Disinformation narratives about Ukraine were most frequently spread through Facebook communities and on YouTube channels.

II. *Disinformation object:* authorities (the President of Ukraine, government, local self-government bodies).

Metanarrative: External governance of Ukraine, The Ukrainian authorities are neo-Nazi, illegitimate and discredited.

Goal: discrediting Ukraine's political authorities; demonstrating the incapability of Ukrainian authorities and the divided society; pressure on the political authorities with the purpose of securing certain decisions in the field of politics.

Examples: "Ukraine is a Western puppet"; "The West in general, and the United States in particular, have taken Ukraine under control"; "Soros minions in the authorities are agents of external governance"; "There are certain misunderstandings or tensions between the political leadership of the West and Ukraine";

"Ukrainian government is corrupt and incompetent"; "There is a clear conflict among the Ukrainian authorities"; "The illegal junta in Kyiv"; "Neo-Nazis have seized power in Ukraine"; "Ukrainians need a new government"; "The Ukrainian government capitulates"; "Citizens of Ukraine are dissatisfied with their authorities, which will cause destabilization, mass protests and early elections"; "The President of Ukraine launches repressive measures against pro-Russian media"; "Zelensky's power is weak, he has no control over the situation, and he will be removed when the right time comes"; "Due to decentralization, regions will no longer listen to the weak Kyiv"; "Zelensky won't take the risk of fulfilling his promises as he is controlled by oligarchs and Soros"; "The Ukrainian authorities do not adhere to the Minsk agreements and breaches the arrangements"; "The President acts primarily in the interests of Western Ukraine and ignores Russian-speaking regions."

Dissemination channels: YouTube, Facebook, Twitter, Instagram.

III. *Disinformation object:* The Revolution of Dignity (protesters, public figures, representatives of political parties, volunteers, medics).

Metanarrative: The Revolution of Dignity is an unconstitutional seizure of power by force, an armed coup d'état.

Goal: encouraging a division of society, denying expression of Ukrainians' will.

Examples: Events on Maidan are an armed coup d'état with the support of Western security services that resulted in the loss of Crimea, civil war and deterioration in Ukrainians' quality of life; nationalists stand on Maidan; Struggle of nationalist forces supported by the West; The Revolution of Dignity was supported by the West; The Euromaidan led to Ukraine's break-up in all areas; Ukrainians bit at the Western bait and can't recover from the consequences; The Revolution of Dignity resulted in a financial and economic crisis; The Revolution of Dignity is political mistake, and a wrong vector for Ukraine; The Revolution of Dignity led the country to a dictatorship; The Euromaidan is the victory of Russophobia.

The disinformation narratives are full of myths and stereotypes that demonise Maidan, indulge in negative labelling and call for the Russian-speaking population of Ukraine to "defend themselves". They create an image of the enemy personified in the new Kyiv authorities and the part of the Ukrainian people who strived to dismantle Yanukovych regime.

Dissemination channels: YouTube, Facebook, Twitter, Instagram.

IV. *Disinformation object*: war (military authorities; the Armed Forces of Ukraine, volunteer battalions).

Metanarrative: Russia is an innocent victim; In Ukraine, there is a war / special military operation.

Goals: legitimisation of Russia's full-scale invasion of Ukraine; substantiation of the claim that Russia only acts in self-defence; focusing on Ukraine's defeats and concealing its achievements; discrediting the Ukrainian army; reducing people's trust in the AFU, volunteer battalions, local territorial defence units.

Examples: "Ukraine won't be able to hold out in the war against the Russian Federation and will quickly lose it"; "Ukraine is going to lose the war"; "With the help of the West, Ukraine was going to destroy the Russian Federation"; "Ukraine actually dreams of conquering Europe"; "Foreign mercenaries fight in Ukraine"; "The West made Russia attack Ukraine"; "Ukraine is all Nazi, so they should be denazified"; "We are losing but they don't tell us about it"; "The Ukrainian military poisons water supplies"; "AFU soldiers sold HIMARS to Russians for $800,000"; "Mobilization in Ukraine is conducted with violations of laws"; "Ukrainian nationalists messed up evacuation from Mariupol"; "Ukrainians can't handle weapons / Ukrainian army is not combat-effective"; "Ukrainian military command doesn't take care of their soldiers and puts them at risk"; "The Ukrainian military breach the ceasefire"; "The Ukrainian military use disinformation to conceal their defeats / losses"; "Ukrainian soldiers do not receive the logistical support / compensation they are promised"; "The Ukrainian military are going to blow up a children's hospital in Kharkiv"; "Nationalists have struck at Kakhovka Hydroelectric Plant with Tochka-U missiles"; "Legalization of DPR/LPR"; "The killers of Donbas"; "Ukrainian troops are demoralized and don't want to fight"; "The Ukrainian military shell the people and civil infrastructure in the so-called DPR and LPR and commit genocide in Donbas"; "The Ukrainian army commits war crimes"; "The AFU consists of people with alcohol and drug addictions"; "Territorial defense units are cannon fodder, nobody is actually ready to fight"; "AFU fighters take away food from civilians, while the Russian army feeds the people of the occupied territories"; "The AFU has surrendered"; "The AFU are executing Mariupol residents"; "For the war to finish, Ukraine has to cede its territories"; "Azov are Nazis (they are shooting at population centers)."

Dissemination channels: YouTube, Facebook, Twitter, Instagram.

V. *The object of disinformation*: international organisations (NATO, UN, EU, WEU, IMF, CE, OSCE, IAEA).

Metanarrative: European integration is a crime against the Ukrainian people; the EU is weak and divided.

Goal: creating distrust of the Ukrainian society for the European Union and NATO, the Alliance member states, international partners, their diabolisation and discreditation.

Examples: "The West exploits Ukraine and is destructive to the world order"; "A NATO general is arrested"; "NATO tries to surround Russia with the help of Ukrainians, and attempts to take over the world leadership using the military approach"; "The USA are rewriting the history of Ukraine and World War Two"; "The USA lobbied the ban of Russian social media"; "The USA is an aggressor which tries to unleash war in Donbas, using Ukraine as a tool"; "Poland wants to annex part of Ukraine"; "NATO countries have actually occupied Ukraine"; "The International Monetary Fund is controlling Ukraine"; "There are thousands of U.S. and Allied states military servicemen in the territory of Ukraine"; "The downfall of Western civilization is inevitable"; "The NATO response to the Russian-Ukrainian crisis and the corresponding reinforcement of the Allied states in Central and Eastern Europe are a violation of the Alliance's international obligations"; "The West is tired of Ukraine"; "NATO thinks in Cold War era categories"; "The Alliance planned to deploy ships and missiles in Crimea"; "All decisions in NATO are made by the United States"; "NATO is surrounding Russia".

Dissemination channels: YouTube, Facebook, Twitter, Instagram.

By way of generalisation, we can conclude that political disinformation narratives are tailored for: (1) citizens of Ukraine (except for the population of the temporarily occupied territories of Donetsk and Luhansk regions) and aimed at provoking feelings of disappointment and hopelessness regarding the "Western direction" among the patriotic-spirited members of the public; at reinforcing the desire of the pro-Russian part of the population to reestablish the dialogue with the Russian Federation on the conditions dictated by the Kremlin; at returning the territories occupied by the Russian Federation (except for Crimea and Donbas) to Ukraine but on the Kremlin's conditions; at Ukraine's abandonment of Crimea at the change of the current government and political elite in favour of pro-Russian authorities; (2) residents of the temporarily occupied territories of Donetsk and Luhansk regions and aimed at provoking the feeling that they won't be able to return to Ukraine as long as it sticks to the pro-Western course; at provoking the feeling of hatred for the current government of Ukraine, NATO, the European Union, and the West in general; at blaming Ukraine for all social and economic problems; at creating a positive image of the Russian Federation and its leadership; at facilitating integration of the occupied territories into the linguistic, information and cultural space of the Russian Federation; at creating preconditions for returning the said regions to Ukraine after the government becomes a pro-Russian one; (3) citizens of the European Union, the USA and international partners of Ukraine, and aimed at forming the image of Ukraine as a "failed state"; at creating an anti-Ukrainian discourse in Western media; at forming an information trend proving Ukraine's historical appurtenance to the so-called "Russian world"; at impelling Western governments to weaken or revoke anti-Russian sanctions; at propagating the pro-Russian worldview; at removing the issue of occupation of Ukrainian territories from the socio-political agenda.

Social disinformation narratives. Having monitored social media, one can conclude that social disinformation narratives aimed at discreditation of internally displaced persons with consideration of their social and political status, age, peculiarities of national mentality, and religious views are actively generated online; such narratives also strive to discredit the Ukrainian health care system, as well as initiators of the medical reform etc. There is a series of disinformation narratives visible about internally displaced persons. In particular, the metanarrative "Due to nationalistic views dominating Western Ukraine, internally displaced persons have to leave their new homes" is actively spread in social media (Facebook, YouTube, Twitter), with the goal of discrimination based on territorial and worldview differences. Other examples may include such sub-narratives: "Refugees hate Western Ukraine because there are no combat actions"; "People in Western Ukraine are opposed to refugees who may take away their jobs"; "The West is unfriendly to refugees"; "All residents of Donbas are looking forward for Russians to come."

In the course of monitoring, disinformation narratives are apparent concerning the country's health care system / medical reform. The metanarrative "The health care system reform is ineffective", which is intended to disqualify initiators of the medical reform and block its implementation, was spread throughout various communication platforms. It was accompanied with the following derived sub-narratives: "Suprun has liquidated standardization of medical aid"; "In the country of chronic epidemics, they have liquidated internship training for infectious diseases"; "Soros minions are interfering in health care"; "The health care reform has failed"; "Introduction of electronic medical records is dangerous for Ukrainians." Social media were spreading destructive criticism against the former Minister of Health Uliana Suprun, who was blamed for the alleged failure of the medical reform and for her desire to "eliminate the Ukrainian nation".

Results of the research indicate that disinformation narratives concerning COVID-19 account for a large volume of social media content. The metanarratives "Coronavirus is a fiction" and "Vaccinations are dangerous to health" have given rise to a number of sub-narratives, in particular: "Coronavirus is faked by politicians"; "Vaccination kills"; "Vaccines are of poor quality"; "Vaccination against COVID-19 will be used to control the population"; "Ukraine is unable to protect its people against the pandemic"; "The government wants to give everyone a poor quality vaccine"; "Politicians / officials don't tell us all the truth about the vaccine"; "Ukraine is supposed to use the Russian Sputnik vaccine but the external governance won't allow it"; "Russia will supply vaccines to the DPR / LPR, Moldova and Transnistria"; "The Kremlin protects its people, but Ukraine will be left without any vaccines"; "Ukrainians will be used to test the vaccine"; "New strains of coronavirus are a part of the process to modernize biological weapons which are being tested on Ukrainians in secret laboratories in Ukraine under U.S. control"; "Bill Gates wants to implants chips to people's bodies to control them"; "The USA planned to use Ukrainian birds and mosquitoes to transmit diseases and destroy the Slavic ethnos"; "Coronavirus is spread with 5G technology"; "Coronavirus is the attempt of world governments to control the population and reduce its numbers" etc.

The goals of disinformation narratives concerning COVID-19: denying the existence of coronavirus or belittling its danger; initiating protests and mass riots supporting refusal of vaccination; negative effects of vaccination; making havoc in the society; manipulating people's opinion; stirring up hatred or provoking confrontations.

By the topics involved, disinformation narratives concerning coronavirus may be usefully divided into the following groups: 1) narratives about people ill with coronavirus: their number, personalities, correlations with race, gender and other characteristics ("Everything depends on the skin color: scientists surprised everyone with their new discoveries about coronavirus", "People with a certain blood group are at greater risk to catch COVID-19"); 2) narratives concerning the virus: its origins ("China is where COVID-19 originated", "Coronavirus is a development of American / Chinese scientists and it was created in a laboratory", "COVID-19 is a weapon created by NATO"), causes for its emergence ("The development of coronavirus was sponsored by the Bill & Melinda Gates Foundation"), ways and means of its transmission ("Snakes are intermediate hosts of SARS-CoV-2", "The virus is transmitted by mail"), structure, forms and size; 3) narratives containing pseudo medical advice: ways of treating and combating the illness, medicines and folk remedies, self-treatment ("The prevention method is to keep your throat moist", "Advice from a young doctor Yuriy Klimov from Wuhan"), ways of self-diagnosing the virus; 4) anti-vaccination narratives ("Vaccines can't save you from death", "Vaccines cause mutations", "CoviShield is of poor quality, so even African countries refused to take it"); 5) disinformation about social and economic consequences for the world order or certain region ("A few thousands of death from coronavirus triggered the global chaos").

Their authors used various forms of presenting the disinformation content, including faked screenshots, videos, edited photos, documents, and stories of new coronavirus cases. The platforms most frequently engaged in spreading disinformation include Facebook (such groups as STOP Fake Pandemic, "Vaccination: Free Choice"), YouTube, and Instagram. Active agents who spread disinformation narratives concerning coronavirus on YouTube channels were bloggers Ostap Stakhiv, Anton Gura and Petro Sologub, who portrayed themselves as human rights activists.

Historical Disinformation Narratives

The object of disinformation includes Ukrainians, national minorities, and post-Soviet countries. The metanarrative "Ukraine is an artificial state that only emerged in the twentieth century thanks to the USSR" / "Ukraine is an artificially created state construct" / "Ukraine doesn't have its own history" is broadcast through various communication channels.

The goals of historical disinformation narratives include the following: falsification of Ukrainian history; denial of Ukraine's national identity; propagation of the myth of common history; glorification of the Soviet Union and its role; provocation of national and ethnic strife; support for separatist tendencies.

Let us adduce examples of derived sub-narratives: "Ukrainians are an eternal and inseparable part of the triune Russian nation"; "Common Soviet history"; "Little Russians lacking their own language and history"; "Historically, Ukraine was on the periphery of the former Russian Empire"; "Ukraine emerged as a result of Russian Bolsheviks' and Lenin's efforts"; "The USSR is a cradleland of nations"; "Ukrainians and Russians are a single brotherly nation"; "The Soviet Union was a superpower"; "Donbas is Russian"; "Crimea is Russian"; "Oppression of the Russian language in Ukraine"; "The Russian minority is subject to oppression in Ukraine"; "Discrimination and humiliation of the Russian national minority due to the Law on Indigenous Peoples"; "Russia has the right to combat fascism in other countries"; "In the USSR, people had a guaranteed pension"; "There was no Ukraine in the south, but there was the so-called Novorossiya"; "The Baltic states should be thankful for being allowed to secede from the Soviet Union"; "Russia has given the Baltic states the freedom which they dreamed of."

Cultural Disinformation Narratives

The object of disinformation is Ukrainian culture (artists, cultural products, traditions). In the course of the analysis, we have distinguished disinformation narratives related to the authenticity of the Ukrainian culture: "Ukrainian culture is poor"; "Russia is a civilization-building cultural center around which other countries should unite"; "Ukrainian culture is provincial"; Ukraine steals Cossack traditions from Russia"; "Shevchenko is a Russian-Ukrainian poet" etc. Goals of disinformation: marginalisation of Ukraine's culture, bringing down Ukrainian cultural values.

Religious Disinformation Narratives

The object of disinformation is the Orthodox Church of Ukraine. The metanarrative of "the Ukrainian Orthodox Church (Moscow Patriarchate) being the only canonical church in Ukraine, and the Orthodox Church of Ukraine being just a schismatic organization" is actively spread in social media (Facebook, YouTube, Twitter). Its goal is to discredit the Orthodox Church of Ukraine (OCU) and provoke religious conflict. It is accompanied with such derived sub-narratives: "The OCU is created by politicians to struggle against Russia"; "The Orthodox Church of Ukraine is not a canonical one"; "The OCU is under external governance – it is dependent on the Greek church, so there is no autocephaly"; "Zelensky failed to restore justice in regard to the Ukrainian Orthodox Church (Moscow Patriarchate) and to stop religious conflicts because he (and the West) benefits from splitting the Ukrainian Orthodox community"; "The Ukrainian Orthodox Church (Moscow Patriarchate) is beyond politics and independent from Moscow".

By request of the civil organisation Detektor Media, Kyiv International Institute of Sociology (KIIS) has carried out a survey to identify disinformation narratives in the Ukrainian segment of social media and explore Ukrainians' reactions to social challenges. They have analysed over 23 million posts from eight target areas in such social media as Facebook, YouTube, Instagram, and Twitter.

The survey studied disinformation narratives concerning "external governance" and the coronavirus pandemic. According to the survey data, residents in southern and eastern regions of Ukraine tend to believe in "external governance": 69 % of the respondents think that cooperation with the International Monetary Fund causes Ukraine to be controlled from abroad; 20 % believe that this cooperation gives Ukraine the aid it needs; 58 % adhere to the opinion that since 2014 the most important decisions in Ukraine have been approved under the influence of Western countries; 42 % see the anti-corruption infrastructure as a tool for external governance of Ukraine, while 18 % consider it an element of combating corruption; 24 % trust that important decisions are made by Ukrainian politicians [11]. Over the period of this monitoring, the highest number of posts supporting the narrative of "external governance" was recorded in social media located in Donetsk and Odesa regions (see Fig. 3).

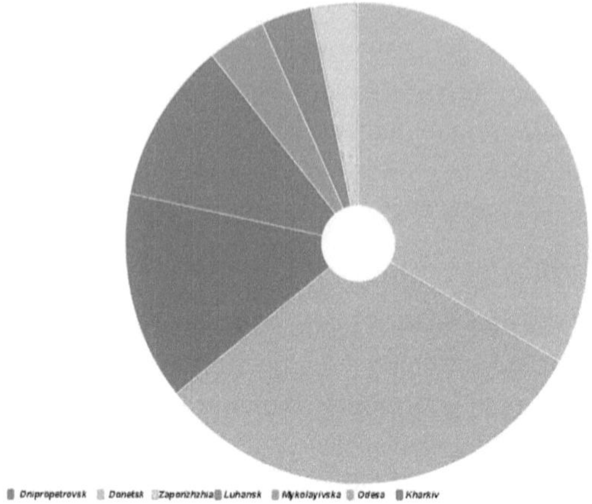

Fig. 3. The audience reception of disinformation narrative of "external governance"

As to the disinformation narratives about COVID-19, the results of this sociological research indicate that 54 % of the population in southern and eastern regions adhere to the opinion of the artificial origin of the coronavirus, while 25 % believe it is of natural origin. Additionally, 47.5 % of the respondents claim that there is no cure for coronavirus, and 18 % chose the option "the cure for coronavirus does exist, but this information is concealed". Similarly, disinformation narratives concerning the measures applied to contain the pandemic are quite widespread. For instance, 49 % of the respondents find quarantine and other pandemic-related restrictions to be essential, while 26 % think that the danger of coronavirus is exaggerated, so the quarantine and restrictive measures are not necessary [14].

To learn the respondents' opinion as to Russian disinformation and determine the rate of its effectiveness, the 2020 data was compared with the results received by Kyiv International Institute of Sociology in 2015 [14]. The assessment scale was marked from 0 to 100 points, where 0 indicates minimum disposition to believe Russian disinformation, and 100 indicates maximum disposition. The monitoring showed that the average figure for eight eastern and southern regions over the five years remained at the level of 37 points. In 2015, about 19 % of the respondents had from 60 to 100 points, that is, were disposed to believe Russian propaganda; whereas now this figure has decreased to 13 %. The share of respondents who scored from 0 to 40 points (immune or non-disposed) has only grown slightly to 62 %, whereas in 2015 this figure was 60 % (see Fig. 4).

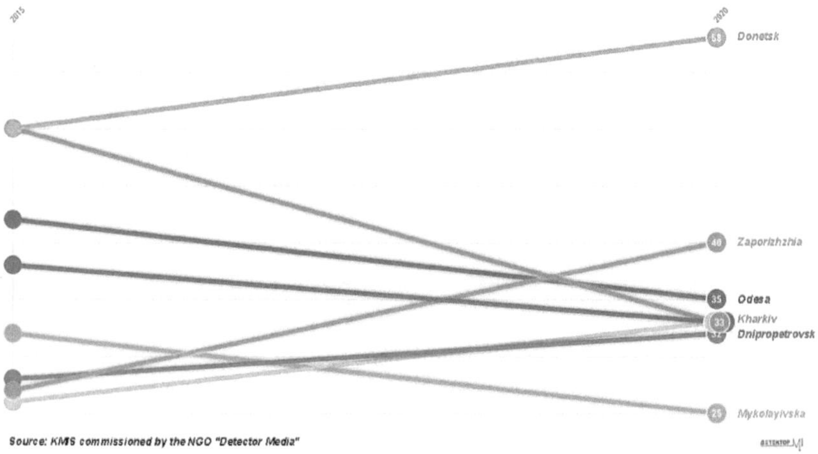

Fig. 4. Disinformation effectiveness ratio

Conclusions

Summing up, the main threats resulting from dissemination of disinformation narratives in social media can be distinguished as follows: the large number and variety of their forms; interference in social-political processes and transformations; latency and insidiousness of disinformation strategies. In the period of social transformations, disinformation texts create a discourse which envisages spreading information chaos, demoralisation and demotivation of public opinion, promotion of total scepticism, cynical mind, and apocalyptic moods.

In the course of content analysis, it was discovered that pro-Russian (63 %) and anti-Western (37 %) disinformation narratives are intensively generated in social media. Classified as pro-Russian are the narratives related to discreditation of the Ukrainian state, the current government, the army, the people etc.; and classified as anti-Western are the narratives aimed at discreditation of the USA and the European countries, as well as international organisations. In order to distort and transform social reality, authors of disinformation narratives use various techniques, including contrast (Ukraine/Novorossiya/Russia, Eastern Ukraine/Western Ukraine, Russia/the USA/NATO/the EU, friend/foe, Catholic/Orthodox Ukrainian "fascists"/"liberators"); stereotyping (creation of stereotypes for "Banderite", "enemy of the people", "Kyiv junta"); mythmaking (cultivating myths of "brotherly nations", "the great Victory", Cold-War fear of the West, fear of Western values etc.); emotional-semantic technologies with negative connotations (intimidation, propagation of fear and hatred, threats); likening (to Soviet or Russian things); discreditation (destroying an opponent's image or distorting the image of a certain event, phenomenon or process); idealistic speculations (nostalgia for the USSR, "ideals of Communism") etc.

The following lines of activity may be suggested to counter disinformation narratives: (1) monitoring of social media by specialised groups of disinformation countering experts: documenting disinformation texts, identifying communities that generate such narratives, and drawing analytical case files on these; (2) awareness: using social media to inform people on their actions in case they face disinformation narratives; (3) media literacy and critical thinking: checking sources of information; comparing materials from various / alternative sources; carefully considering facts and taking independent decisions; (4) involvement of opinion leaders: referring to well-known authorities from various fields of life; (5) active citizenship: responding to disinformation materials posted in social media; thought-out and balanced discussion with their authors; using feedback options.

The leading mechanism in countering disinformation is represented by various fact-checking projects and civic organisations whose activities are aimed at identification, examination and refutation of false content. With grant support from government agencies and international foundations, Ukraine has created a series of projects and organisations engaged in monitoring of information influences and analysis of disinformation texts. For example, the Detektor Media civic organisation, the Center for Analytics and Investigation civic organisation, the Center for Countering Disinformation at the National Security and Defense Council of Ukraine, the Center for Strategic Communications and Information Security under the Ministry of Culture and Information Policy of Ukraine, the Institute for Mass Information, and such projects as BezBrekhni, IREX Learn to Discern: Media Literacy, VoxCheck, StopFake, Nadiyno etc.

Therefore, to counter disinformation effectively in the national information environment, it is necessary to implement a cross-institutional approach, as well as use such methods as fact-checking, monitoring, cooperation with Internet platforms, media and leading fact-checking organisations, and promote positive narratives through certain communication policies.

References

[1] Bilousenko O. Yak Yevropa vidpovidaie na dezinformatsiiu, vodnochas zakhyshchaiuchy pravo na vyrazhennia pohliadiv. https://detector.media/infospace/article/185095/2021-02-21-yak-ievropa-vidpovidaie-na-dezinformatsiyu-vodnochas-zakhyshchayuchy-pravo-na-vyrazhennya-poglyadiv/.

[2] "Detektor media": khto i dlia choho rozpovsiudzhuie prorosiiski naratyvy v ukrainskomu mediaprostori. https://petropavlivka.city/articles/157552/detektor-media-hto-i-dlya-chogo-rozpovsyudzhuye-prorosijski-narativi-v-ukrainskomu-mediaprostori.

[3] Dezinformatsiia v sotsialnykh merezhakh. https://www.oporaua.org/news/vybory/disinformation/23286-znaiete-shcho-e-6-dezinformatsiia-v-sotsialnikh-merezhakh.

[4] European Commission. A Multidimensional Approach to Disinformation. Report of the Independent High-Level Group on Fake News and Online Disinformation. Luxembourg, Europe: Publications Office of the European Union, 2018a.

[5] Fallis D. What is disinformation? Library trends, 2015. № 63.3. pp. 401–426.

[6] Koretska I. O. Feiky yak instrument vplyvu na vybory. Analitychna dopovid. 2020.

[7] Metodolohiia monitorynhu feikiv ta rosiiskoi dezinformatsii v ukrainskykh onlain-media. Instytut masovoi informatsii. 2019. https://imi.org.ua/monitorings/metodolohiia-monitorynhu-feykiv-ta-rosiys-koi-dezinformatsii-v-ukrains-kykh-onlayn-media-i28321.

[8] Oxford English Dictionary. http://www.oed.com/view/Entry/54579?redirectedFrom=disinformation.

[9] Pocheptsov H. H. (Dez)ynformatsyia. K.: PALYVODA A. V., 2019. pp. 198–220. http://csef.ru/media/articles/9072/11714.pdf.

[10] Po toi bik ekrana. Analiz mediaspozhyvannia ta dezinformatsii v ukrainskomu informatsiinomu seredovyshchi. HO "Detektor media", 2021. https://detector.media/doc/images/news/archive/2021/188114/On_the_other_side_DM_final_UKR_WEB.pdf.

[11] Reitynh populiarnykh saitiv za berezen 2022. https://tns-ua.com/news/reyting-populyarnih-saytiv-za-berezen-2022.

[12] Shcho take dezinformatsiia, mizinformatsiia, propahanda? Prometheus. 2021. https://prometheus.org.ua/.

[13] Yakisna dezinformatsiia pidtverdzhuie vashi pohliady, i neiu khochetsia podilytysia. Pereklad iz amerykanskoho vydannia. The Rolling Stone. TEXTY.ORG.UA. 2019. https://texty.org.ua/articles/98579/Jakisna_dezinformacija_pidtverdzhuje_vashi_poglady_i_neju-98579/.

[14] Zykun N. I., Bessarab A. O., Ponomarenko L. H. Naratyv yak kontentna osnova mizhnarodnykh stratehichnykh komunikatsii. Derzhava i rehiony. Seriia: Sotsialni komunikatsii, 2020. 1 (41). pp. 4–11. https://doi.org/10.32840/cpu2219-8741/2020.1(41).1.

Jakub Kruk

Political Narratives in Media: Analysis of Sentiments in Statements of Politicians and Their Impact on the Public Opinion

Introduction

At present, humanity is in the information age, in which access to information is no longer only the privilege of the most well-read and well-educated, as was the case before the invention of print and universal access to education. At present, access to information is a part of daily life and now we all have the right to access reliable information. History shows that not all rights granted to society are granted once and for all. Effective protection of the freedom of speech depends above all on the attitude of the citizens, in particular in enforcing the observance of the applicable rights by the authorities. On the other hand, the media are expected to observe appropriate journalistic standards, but when the situation requires it, civil society should stand in their defence when the existence and independence of the media is jeopardised. Without a democratic society, there will never be a free press, but without free press there can be no democratic society.

In recent years, the Polish media and political scene have been continuously subject to a deep process of division into two antagonistic camps – one connected with the governing coalition and another connected with opposition groups. This process is commonly felt; however, it is relatively rarely presented in analytical terms. This study is an attempt to describe this phenomenon, which will be made possible by the application of computer science tools from the widely understood and continually developing domain of big data in the analysis.

The results presented are the result of a several weeks-long process of collection, processing and analysis of Polish media texts. The result of these analyses opens the way to discussion on the state and level of the political debate in Poland and motivates further research on the forms of manipulation and disinformation present in our environment and on the meaning of rhetoric in modern times and political and social circumstances.

The main and most important objective of this study is to verify the hypothesis that the narrative conducted by Polish politicians in the media space influences Polish public opinion. At the same time, the purpose of this study is to answer the question to what extent, in the studied period, the political narrative of the individual political parties changed depending on the political and social topics that appeared or that they provoked. In addition, the following analysis will make it possible to study the dependencies between the message of individual media outlets and the narrative of individual political parties that can be generally associated with them.

Society and Media

The relationship between the media and the audience is continually evolving before our eyes. New parallel forms of communication and information communication, such as television and the Internet, effectively distort the reality known from the 20th century. The press, both traditional and Internet, is forced to compete for audiences, which is often reflected in the quality of the materials produced. As a rule, fighting for survival, the media have become susceptible to political or business influences. Over time, forced by situation, they combine into larger media groups, which in the long run kills their diversity, because the least profitable outlets are closed.

At present, social media play a particularly important role (on an unprecedented scale) in communicating information. Every year, the mass media audience worldwide is growing exponentially. In 2018–2020, the number of Internet users worldwide increased by ca. 300M people. The COVID-19 pandemic period accelerated this process even more – in 2021, the number of Internet users increased by 400M persons [STATISTA]. Providing all those users with access to verified information and news is already a global challenge, which grows even more difficult year by year, decade by decade.

Undoubtedly, this immense scale and continually increasing audience is constantly exposed to all methods of manipulation, disinformation or propaganda that we know today and that we will know in the future. This threat very often results from the people not being aware of the exposure to any of the abovementioned phenomena. This results from the lack of proper knowledge, habits or using standard thought schemes, i.e. "How can a newspaper or TV station lie?".

Propaganda

Inattentive (or unaware) recipients of the contents communicated in the media may have a hard time noticing the line between rhetoric and propaganda, even though they have accompanied humankind since the time immemorial.

The first foundations for the scientific approach to the art of speaking were written already in Ancient Greece, in one of the most known works of Aristotle, Rhetoric, in which the philosopher distinguished three types of speech – deliberative (political), forensic and epideictic (see Ref. [1], p. 43). The practical dimension of Aristotle's Rhetoric found application already in antiquity, during the rule of Alexander the Great, the philosopher's student, only to be improved and developed by the authorities of the Roman Empire. The methods of gaining social support were mastered to perfection by organising games, demonstrations or processions after successful military campaigns. Over the following centuries, these principles were the grounds for the development of the modern rhetorical concepts known to us today.

Propaganda, however, compared to the art of rhetoric, is often very primitive in its form and content, while in "proper" hands it may become quite refined. Examples of its use and impact may be found in the history of practically all great civilisations of antiquity, that is, Ancient Egypt, Assyria, and Babylonia (see Ref. [2], p. 63). The examples include early hieroglyphic writings from the Nile and Mesopotamia, in which the history of that time is presented favourably from the point of view of the ruling classes by praising them (see Ref. [3], p. 17). Propaganda practices have always been effectively used by the representatives of religious communities and great religious systems, which triggered the development and thus the application of effective persuasion of societies. Building a consolidated religious community was by design based on the assumption that all worshippers had to have a personal conviction that they want to belong to it and believe in common values, of course except for the "conversion" of the infidels by force. In late antiquity and the early Middle Ages, the main means for the dissemination of propaganda contents were travelling priests or temples. In addition, the institutionalisation of religious structures helped forming a uniform message, even in the most remote parts of the world know to the people of that time. An example could be the successful "campaign" of Pope Urban II, who promoted mass participation in the crusades. Looking from today's perspective, this could be undoubtedly referred to as the first "global propaganda campaign". It touched its audience by invoking their emotions, common Christian values and the potential temporal and material gains after expelling the Muslims from the Holy Land. It resulted in drawing nearly all European countries in an almost

200-year conflict with the Arab world (crusades against Muslims were organised in 1095–1291).

For this reason, among other things, the invention of print by Johannes Gutenberg in 1450 should be considered as one of the greatest breakthrough moments in human history. For example, this invention was a fundamental driver for the Reformation, thanks to the good use of the new communication channel by Martin Luther. Print enabled above all breaking the information monopoly held until that time by the Catholic Church and the lay rulers in Europe. It directly translated to the development of education, in particular the mathematical and social sciences, because print also accelerated the information exchange between people and scientific institutions.

The blossoming of the press, which followed in the 18th century, modernised the form, in which information was transferred, and this developed form has been known until today, as we have contact with journalism in the broad sense on a daily basis. At the same time, Gutenberg's invention enabled propaganda of a mass character. The events that may be considered as the subsequent milestones in the development of political communication was the American Revolutionary War and the French Revolution, which gave birth to modern democracies. Starting with these events, political propaganda began to take two faces. The first was connected with the development of state propaganda. State bodies and institutions refined the methods of influencing domestic and international public opinion. The second face showed in the first political rivalries between the forming political parties and organisation, referred to as election propaganda. Its purpose was to convince potential voters through classical persuasion methods and the relatively new method, the partisan press (see Ref. [2], pp. 65–70).

The term "propaganda" became commonly used in the early 20th century, due to the First World War, with reference to the "persuasion tactics", which were an indispensable piece of combat, next to military operations. Later this term became commonly used in the context of totalitarian states (see Ref. [3], p. 17). This may be the reason why this term does not evoke positive associations and triggers negative emotions. It is often synonymous to manipulation or "brainwashing".

According to the definition in Słownik Języka Polskiego PWN, propaganda is "wilfully influencing populations and individuals, aiming at gaining supporters and allies, instilling desirable convictions and causing specific endeavours and behaviours. Usually, the purpose is to promote (impose upon others) permanent attitudes; then propaganda is a channel of indoctrination" [PWN]. To complement this definition, it should be added that the effectiveness of propaganda assumes that its creators effectively use images, slogans and symbols to present exactly the point of view that the audience will be inclined to willingly accept as their own (see Ref. [3], p. 17). For this very reason, it is a crucial component of order in authoritarian and totalitarian systems, in which the purpose of the authorities is to make the audience dependent on them, while making them unable to verify the incoming information, because then the effectiveness of such messages is even greater and thus more dangerous. Globally, it may be used to spread ideas, religions or political thought, as was the case during the Cold War, when two major economic, political and military blocs clashed, one led by the United States of America and another under the aegis of the Soviet Union.

"Total" propaganda is also a constant component of the information wars, which intensified the most during full-scale armed conflicts, starting with the Second World War, through the war in Yugoslavia in 1990s, to the war in Ukraine started by Vladimir Putin in February 2022. To compare the rhetoric and propaganda, it is worth quoting the article by Elżbieta Pieniążek-Niemczuk of 2016, in which she presented the difference between the two terms: "Rhetoric is aimed at persuading through the beauty of language, while propaganda frequently includes a negative manipulation of language" (see Ref. [4], p. 70).

This difference is clearly apparent. The two methods of communicating information are contrary to one another. Rhetoric assumed that a subject is addressed under conditions leading to a discussion, in which the speaker is exposed to confrontation with other speakers, which could be both the audience and other rivals, for example, the opposing team during an Oxford debate, candidates in elections or representatives of other political parties. The purpose of rhetoric is to influence the emotions of the audience (pathos) using rational argumentation (logos) (see Ref. [1], p. 43). In turn, propaganda is intended only to provoke emotions, while skipping the "reasoning". The creators of propaganda focus on selecting facts and arguments to obtain what to them is the optimal effect. This is achieved at the expense of rational argumentation, because the propagandist who decides to consciously deny the relevant facts to disrupt vigilance and awareness, often disturbs the cause-and-effect flow of their own reasoning (see Ref. [4], p. 70). In effect, critical assessment of all propaganda content is their greatest enemy, because it is relatively easy to confront them with the actual state of affairs. It is often enough to merely refer to the facts and events omitted. Every more attentive propaganda recipient can debunk it, but they must be capable of doing so. This is what we can do thanks to free media and freedom of speech.

Fake News, Disinformation and Post-truth

The marketplace of free ideas defined by John Milton is a place of competing perspectives and different arguments. This is also where fake news is disseminated effectively, which often distort the image of the surrounding reality, as they are intended to. Ironically, the creators of fake news and the people wilfully disseminating them invoke their right to free speech in order to seemingly justify their actions and blur the scale of the social harm they cause. Undoubtedly, they move beyond the moral lines of freedom of speech.

The world of science and journalism incessantly strives to define the notion of fake news. The terms "false news" or "fake news" began to appear in the 19th century in newspaper articles of that time, in which these terms were mainly used to describe the plague of made-up and sensational news in other, competing newspapers [5]. Today, this phenomenon is even more refined in the form and more problematic, considering the scale of its occurrence.

The review of the definitions available makes it possible to gain a broader perspective. The definitions may be shorter or more detailed, which should by no means be perceived as a defect. The objective should be to unify and simplify as much as possible, so as to make it understood and available as universally as possible. The victims of false information are most often people with less knowledge of the practices used in traditional media or social media, for whom many terms are not explained clearly enough. However, any attempt to describe the fake news phenomenon as precisely as possible should not imply strict formulation in a single finite definition.

The definition of fake news should be considered as an open definition that evolves as continual changes occur in the information space, society and access to technology. None of the existing definitions (in accordance with the open notion of fake news) can be accepted in their entirety and considered as final.

The scholars studying and monitoring fake news increasingly often replace that term with the term "disinformation". However, the definition of disinformation is somewhat more specific. It is very often used as component of a wider, planned undertaking, of which the purpose is to manipulate the selected entity as part of the international information combat. The purpose of disinformation is to indoctrinate and destabilise the enemy societies, in which the role of the disinformer is played not only by armies and special services but also diplomats, businesspersons and journalists. A perfect example of the practices of disinformation is the *modus operandi* of the Russian Federation. Russia is a state in which the disinformation process is institutionalised at the domestic level (TV station Russia-1 [*Russian: Россия 1*]) and the international level (the worldwide Russia Today).

This *modus operandi* goes far beyond propaganda itself, because it is not restricted to merely creating content following the line of Kremlin decision-makers. Russia provokes or initiates various destabilising events or protests, like anti-vaccine and separatist movements, funds organisations and news portals or creates troll farms [6]. The most evident examples were the anti-vaccination group social media profiles operating in Poland, which a few days before the start of the Russian invasion of Ukraine on 24 February 2022 to a large extent turned into websites sharing anti-Ukrainian and pro-Russian content. Several hours before the outbreak of war, the number of defamatory mentions of Ukraine and its inhabitants and pro-Putin posts increased by more than 500 %, while 80 % [7] of all such entries came from pages that had shared anti-vaccination content earlier. Such data clearly demonstrates that the disinformation mechanism designed by the Russian Federation is well-coordinated and centrally controlled – in short, institutionalised.

Another important term, which is important to remember, is "post-truth". Its first use is attributed to Steve Tesich, who used this term in 1992 to describe the phenomenon that appeared in American society during the Watergate scandal. A surprising attitude was noticed among some Americans, who were unwilling to follow news on the scandal, because the reports made then President Richard Nixon look bad. In effect, publishers in the United States restricted the published contents describing the scandal due to low interest among readers. Katherine Connor Martin, the editor of Oxford's U.S. Dictionaries, defined post-truth as circumstances, in which objective facts have less impact on the formation of public opinion than the personal convictions and emotions accompanying them. The term itself was recognised as the 2016 word of the year by the Oxford English Dictionary [8].

The term post-truth is sometimes considered a synonym of disinformation, which is not entirely correct. The two phenomena are similar, but upon closer examination they differ. The most important difference is that disinformation is often strongly connected with propaganda and often pretends to be the only source of information, often imposed by force, whereas post-truth operates symbiotically next to verified information from certain sources.

The post-truth may be regarded as smoke screen released over reliable media contents, because instead of imposing new subjects and creating the whole discussion (as is often the case for propaganda), it "preys" on topics that are already present in media space (see Ref. [9], pp. 176–178).

Modern Strategies of Political Communication

In its principles, building the political message of a political party or state or organisation does not differ from planning the marketing strategy of companies under free-market conditions. In both cases, the individual steps of the strategic management process must be completed, which includes formulating the vision, mission and area of operation of the organisation and conducting strategic analysis. This is often the reason why the political communication strategy is also referred to as political marketing, leading some to treat it as a separate academic discipline.

Political marketing assumes the application of a wide range of theories and concepts of business marketing and marketing of services and non-profit organisations in politics. Therefore, it operates on the border between marketing and political sciences. According to Stephan C. Henneberg "political marketing is establishing, maintaining and reinforcing long-term relationships with voters in order to achieve social and party interests, considering both individual and organisational objectives by mutual exchange of values and fulfilment of needs and promises" (see Ref. [10], p. 13).

Each organisation must define its place in space, its strengths and weaknesses, considering both its actual position (diagnostic level) and the opportunities and potential of the organisation in future (forecast level). Such analysis constitutes the grounds for planning the current tasks, which the satisfactory performance of a broader strategic plan depends. However, despite the significant similarities in the management philosophy and strategic planning procedures between enterprises and political parties, business planning methods cannot be fully applied to the latter. Due to the multitude of external factors influencing the operation of each political party, from the point of view of management, it seems justified to refer only to "strategic thought", rather than specific model solutions (see Ref. [2], pp. 33–41).

The key utility for building political messaging are social media, today constituting the main, but not the only, channel of communication with the voters.

Five political strategy development phases may be distinguished, which are used by politicians and their organisations. The first is the "pre-medial" political strategy phase, of which the planning is based on the existing experience and intuition of organisers. The second phase is referred to as the early mediatisation phase. It involves gradually increasing the importance of the impact of the earlier-formulated strategy on the environment with printed texts. The third phase is the institutionalisation of the communication political campaign. In this phase, specialised institutions and organisations responsible for creating and implementing strategies are established. In the fourth phase, the developed media strategies phase, mass media become the main instrument of implementation of political strategies. The final fifth phase is strategy marketisation. In this phase, selective strategies are directed to key segments of the audience, of course by means of all available mass media (see Ref. [2], p. 44).

Media and Politics

The media are commonly referred to as the fourth power or branch of government, next to the legislature, executive and judiciary, constituting another component according to the classic separation of powers. The media exert control over them. On the other hand, they are subject to legislative regulations and judicial rulings. With this peculiar synergy of four components, a society may examine the government and its activities, about which the authorities often do not like to brag in public. Without free media and independent journalists, no one would know, for example, about Watergate in the United States, the Rywin affair or Pegasus in Poland.

From its very beginning, the media was the main arena of political discourse in democratic countries and the source of basic news, which served the information purpose, while interpreting and commenting on the current political events (see Ref. [11], p. 43). Operating this way, the media have a strong impact on public debate, while often imposing subjects or emphasising existing subjects in that debate. Due to these activities, journalists are often regarded as politicised, which is particularly apparent during the sharp political and social conflict currently taking place in Poland. Consequently, attempts to impose stricter regulations on media entities are controversial, as we witnessed in 2021, when the United Right (*Zjednoczona Prawica*) government first proposed and then passed in Parliament the so-called media act (also referred to as "lex-TVN"), of which the objective was to force the American owners of the media group to sell their shares. The governing party politicians expected them to be purchased by a state-owned company. In the end, the law did not enter into force, as it was vetoed by the President, but the tendency of the politicians to tighten the control over the media was clearly visible and very disturbing. Earlier, a similar idea had been implemented in Hungary, where publicly available independent media do not exist at present. In January 2021, Klubrádió, the last independent Hungarian radio channel in that country lost its licence, three years after a hostile takeover of the largest independent journal "Népszabadság", which overtly criticised the government of Prime Minister Victor Orbán [12].

Despite the failure of "lex-TVN", a similar project to purchase and subjugate commercial media by trusted collaborators with the United Right (*Nowa Prawica*) governing coalition was implemented in Poland. In 2020, PKN Orlen purchased the largest owner of regional newspapers in Poland, the Polska Press Group, and the process of altering the content published by the newspapers and magazines owned by PP (Polska Press) may undoubtedly be called a full subjugation of the published contents to the Law and Justice (PiS) party message.

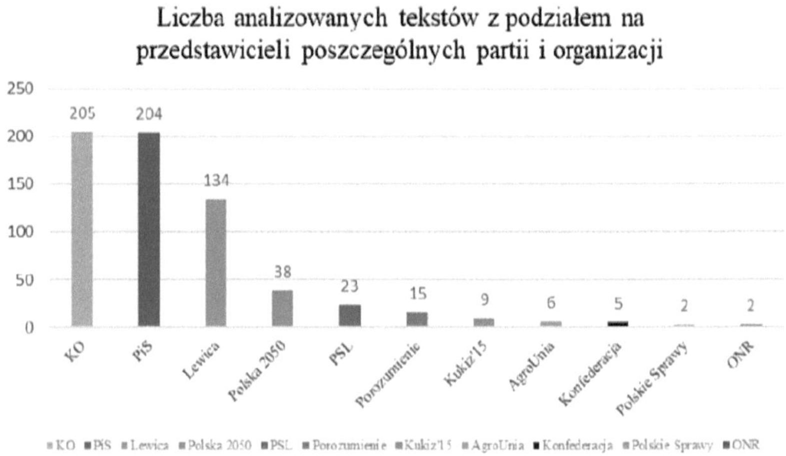

Chart 1. Number of analysed texts divided by representatives of individual party and organisation
(*Source:* original research)

With such important role of journalists in the public debate, when building their image, political parties must strictly subjugate their operation to media logic. Organisations, such as parties, adapt to it both in their organisation and communications, by employing specialised personnel to manage media contacts, appointing official spokespersons (and at the same time appointing them to positions in the highest decision-making structures) and carefully selecting the persons giving interviews to the media. A less formal form of adaptation of political parties and organisations is often having personal relationships with journalists. All this results in the continual overlap of the fields of activity of political entities and the media. The relationships between the media and politicians are complicated, multi-levelled, complex and quite dynamic – this very often results from personal relationships or global politics of the individual parties and their sympathies towards the individual television and radio stations, newspapers or individual journalists (see Ref. [11], pp. 50–52).

The Study

The purpose of the study is to analyse the narrative constructed by Polish politicians in the media space and its impact on Polish public opinion. At the same time, the following frequently occurring predominant terms were studied that politicians in Poland use in the media, depending on the political and social circumstances. An additional analysis will enable studying the dependency of the individual media outlets on individual political parties. The data collected for the study come from the texts published in Polish national media (interviews, abstracts and descriptions of interviews, columns – some written by politicians themselves). Based on these texts, it was possible to determine the character of each statement of politicians, parties in entirety and the general messaging of each media outlet.

There is a commonly known manner among politicians, who change their position or narrative depending on the potential benefits or losses. Sometimes they turn it around 180 degrees. This study shows the scale of this phenomenon in Poland over the few months (October 2021–February 2022) preceding the Russian invasion of Ukraine on 24 February 2022. At the same time, by juxtaposing the surveys of the support for major political parties and monthly survey of the Public Opinion Research Centre (CBOS) on the attitude of the Poles towards the current situation in Poland and the most common sentiments, it will be possible to check if the narrative shift had an impact on public opinion.

Collected Media Texts

According to the research assumptions, all texts collected for study had to be published in the period from 1 October 2021 to 23 February 2022. Each of the media texts was sourced directly from the website owned by the given media outlet, so that the sourced interviews are not re-prints, but original texts. Therefore, the source were websites that offer both free and paid access. To collect the data, scripts constructed in R and Python were used as appropriate.

In total, the analysis included 643 texts from ten different media outlets, that is: Gazeta Polska, Krytyka Polityczna, Newsweek, Onet.pl, Polsat, Rzeczpospolita, TVN24, TVP Info, Wprost and Gazeta Wyborcza. The total word count of all media texts was 484,337. After removing the so-called stop words, that is frequently repeated words, such as prepositions or conjunctions, the total word count subject to analysis was exactly 279,000. The highest quantity of texts is related to representatives of two largest political parties: Civic Coalition (*Koalicja Obywatelska*, KO) and the United Right (*Zjednoczona Prawica*, ZP), which constitute more than 63 % of all analysed texts. The second largest group of politicians, whose statements are included in the study, are politicians connected with the Left (*Lewica*) (at present: New Left (*Nowa Lewica*) and Together (*Razem*)), whose statements constitute more than 20 % of all texts. It is clearly visible that in the examined media group, the percentage of statements of politicians of a given party does not reflect the percentage of support for the individual political parties.

Tab. 1 contains much more interesting information on the "publicity" of individual parties. It is clearly visible that the representatives of the former coalition partner of PiS, namely the party Agreement (*Porozumienie*), are practically not existent in the media favouring the politicians of Law and Justice (*Prawo i Sprawiedliwość*) in Gazeta Polska, TVP Info or Wprost. Over nearly five months, no interview with politicians associated with Jarosław Gowin (the leader of Agreement) was released. This may be interpreted as confirmation of the existing mechanism of restricting access of the "rebel" politicians to some of the media connected with the representatives of the governing party.

This may lead to a false conclusion that the media selecting their interlocutors promote mainly three political circles. However, the examination of the individual articles reveals significant differences in the selection of interlocutors by the journalists of almost each of the studied media outlets. This most often results from the current political situation, events or scandals that naturally involve the selected group of political parties or individual politicians. The subjective selection of interlocutors by the journalists due to their (or their publishers') political leanings is very significant. Not all media outlets try to diversify their messaging (Tab. 1).

Tab. 1. Percentage share of representatives of individual political parties and organisation in studied articles

	KO	ZP	Lewica	Polska 2050	PSL	Porozumienie	Kukiz'15	AgroUnia	Konfederacja	Polskie Sprawy	ONR
Gazeta Polska	0,0%	96,1%	0,0%	0,0%	2,0%	0,0%	2,0%	0,0%	0,0%	0,0%	0,0%
Krytyka Polityczna	27,3%	0,0%	60,6%	12,1%	0,0%	0,0%	0,0%	0,0%	0,0%	0,0%	0,0%
Newsweek	58,3%	0,0%	25,0%	0,0%	8,3%	8,3%	0,0%	0,0%	0,0%	0,0%	0,0%
Onet.pl	41,3%	16,0%	22,7%	8,7%	6,0%	3,3%	0,0%	1,3%	0,7%	0,0%	0,0%
Polsat	15,2%	43,5%	21,7%	4,3%	6,5%	2,2%	6,5%	0,0%	0,0%	0,0%	0,0%
Rzeczpospolita	30,4%	20,9%	27,0%	7,8%	3,5%	4,3%	1,7%	2,6%	1,7%	0,0%	0,0%
TVN24	43,8%	25,7%	18,1%	6,7%	2,9%	2,9%	0,0%	0,0%	0,0%	0,0%	0,0%
TVP Info	0,0%	86,2%	3,4%	0,0%	0,0%	0,0%	3,4%	0,0%	0,0%	0,0%	6,9%
Wprost	25,7%	47,1%	15,7%	1,4%	1,4%	0,0%	2,9%	0,0%	2,9%	2,9%	0,0%
Wyborcza	65,6%	6,3%	15,6%	6,3%	3,1%	0,0%	0,0%	3,1%	0,0%	0,0%	0,0%

(*Source:* original research)

The most extreme cases are Gazeta Polska and TVP Info (with regard to data on TVP Info, the information on the guests of the "Gość Wiadomości" and "Kwadrans Polityczny" programmes does not reflect the entire structure of the politicians participating in TVP programmes, because not all broadcasts of the above-mentioned programmes were described in the form of articles on the TVP Info webpages). In both cases, a significant preference for the political options connected with the governing camp and parties supporting it (Kukiz'15) is visible. In Gazeta Polska, for 51 collected texts (interviews and columns by Joanna Lichocka and Ryszard Czarnecki – PiS member of the Parliament and PiS member of the European Parliament, respectively), 49 texts were interviews or texts by politicians connected with Law and Justice (*Prawo i Sprawiedliwość*), Solidarity Poland (*Solidarna Polska*) or Republican Party (*Partia Republikańska*). The situation in the articles collected from the TVP Info is similar. Of 29 described statements, 25 are programmes with the politicians of the United Right *(Zjednoczona Prawica*, ZP) (86.2 %). It should be noted here that this applies only to the texts available online. In reality, the percentage share of ZP politicians in TVP is slightly lower, yet still significantly higher than the share of the representatives of other parties.

The tendency to select mainly politicians from a single side may be seen in practically every examined weekly and daily, to a varied extent. For example, 60 % of the texts in Krytyka Polityczna can be associated with the politicians of the Left (*Lewica*). The situation in Gazeta Wyborcza and Newsweek is similar, as the politicians of the Civic Coalition (*Koalicja Obywatelska*) gave 65 % and 58 %, respectively, of the interviews from 1 October to 23 February. However, all above-mentioned newspapers are owned by private owners, in contrast to TVP, which is a state-owned broadcaster and is required by statute to be apolitical.

The dependence of the media on the representatives of specific parties or subjugation to their messages is a frequently raised subject. To confirm that such dependency exists, the data collected was subject to the Spearman correlation test in the STATISTICA software. The test results are shown in Tab. 2.

The statistically significant correlations between the variables are marked in red. They show how the narrative of the individual media outlets coincide with the narrative of one political party. In case of Gazeta Polska, the high positive correlation of its average sentiment value with the average sentiment value of the politicians of PiS and its coalition partner parties results from the fact that the politicians of this group constitute more than 96 % of its interlocutors. The correlation of 0.827 means that the shift in the narrative of the politicians of ZP to positive results in a positive shift of the narrative of Gazeta Polska and vice versa. This seems to be a logical conclusion, considering the high share of politicians of this party in the structure of interlocutors of GP. Clearly, on this basis, the media outlet may be described as the private information channel of the Law and Justice (*Prawo i Sprawiedliwość*).

Tab. 2. Spearman correlation test results for sentiment values of political parties and media outlets, October 2021–February 2022

Korelacja porządku rang Spearmana (Sentymenty_partie)
BD usuwane parami
Oznaczone wsp. korelacji są istotne z p <,05000

Zmienna	Wyborcza	Wprost	TVP Info	TVN24	Rzeczpospolita	Polsat	Onet pl	Newsweek	Krytyka Polityczna	Gazeta Polska
AgroUnia				-0,500000	1,000000	0,500000	-0,500000		0,500000	-0,500000
KO	0,546569	0,459236	0,391176	0,576510	0,232072	-0,223025	0,415789	-0,127273	0,252836	-0,246617
Konfederacja	0,000000	-0,200000	0,800000	-0,800000	0,700000		0,100000	0,800000	-0,300000	0,600000
Kukiz'15	0,036037	0,227549	0,500000	0,239625	0,754505	0,454545	-0,107143	0,105409	0,115954	-0,419169
Lewica	0,226490	0,362229	-0,232363	0,389046	0,207228	-0,161074	0,221063	0,681818	0,448916	0,228571
ONR										
PiS	-0,002451	0,054696	0,070588	0,029927	0,121400	-0,462571	0,312281	0,109091	-0,236326	0,827068
Polska 2050	-0,037404	0,477448	-0,391608	0,625303	-0,170448	0,172030	0,582143	-0,275454	0,146559	-0,272259
Polskie Sprawy										
Porozumienie	-0,095238	0,261905	0,095238	-0,250000	0,066667	0,214286	-0,183333	0,085714	-0,392857	0,016667
PSL	0,433333	-0,100000	0,381818	-0,412587	0,188811	-0,209091	-0,663636	-0,200000	0,233333	0,018182

(*Source*: original research)

A surprisingly high positive correlation is found for the average sentiment values of the politicians of the Left (*Lewica*) and the media narrative of Newsweek. The strong positive correlation between them is 0.68.

The high impact on the average sentiment value in the individual week, recorded for TVN24, may be explained by the narrative of Civic Coalition (*Koalicja Obywatelska*) and Poland 2050 (*Polska 2050*). Interestingly, this was 43.8 % and 6.7 %, respectively, of the total number of interlocutors of TVN24. The Spearman correlation value for KO and Poland 2050 (*Polska 2050*) are similar, 0.57 for KO and 0.62 for Poland 2050 (*Polska 2050*), respectively.

Another media outlet, of which the narrative was correlated with the politicians of Poland 2050 (*Polska 2050*), was Onet.pl. They constituted 8.7 % of the interlocutors of "Onet" and the positive correlation of the average sentiment value was 0.58.

In turn, it is difficult to explain the strong negative correlation of the PSL narrative and the media narrative of Onet.pl. The politicians of this party appeared in Onet almost as frequently as the collaborators of Szymon Hołownia of Poland 2050 (*Polska 2050*), namely PSL, constituting 6 % of the interlocutors. PSL, like Poland 2050 (*Polska 2050*), is an opposition party. In addition, both parties are similar in their worldview. It is probable that the interviews given by PSL politicians were both relatively positive and negative, which did not take place in case of the representatives of Poland 2050 (*Polska 2050*), because that party presented itself as a "non-establishment" party and criticised both the government of the United Right (*Zjednoczona Prawica*) and its predecessors. PSL, however, as it was part of several governments over the last 30 years, must defend its decisions or recall its own past successes.

In case of the statements of the representative of Kukiz'15 and the media narrative of Rzeczpospolita, the correlation of the average sentiment values is likely random. The interlocutors from that party were only 1.7 % of the total number of interlocutors of that newspaper, so their apparently real impact on its narrative should also be considered as random and non-existent in reality.

Collected Surveys

To determine the impact of the narrative used by politicians, two public opinion surveys by CBOS were collected: "Nastroje społeczne" (Social sentiments) and "Zaufanie do polityków" (Trust in politicians).

These are surveys organised recurrently at monthly intervals. CBOS conducts its surveys on a group of ca. 1,000 citizens, whose PESEL was selected at random from the pool of numbers delivered by the Ministry of Digitalisation. The monthly "Nastroje społeczne" (Social sentiments) survey addresses four subject areas: progress of the general situation in the country, political situation, economic situation, personal situation and material living standards of the respondents. Every question is asked in parallel with a question about the political preferences of the respondents, which enabled drawing a plot illustrating the social sentiments of Poles from October 2021 to February 2022 (the survey of February did not include the data of the voters of the Left (*Lewica*)).

In addition, to complement the full image of the political and social situation in Poland, surveys of the support for political parties in Poland were collected. They were collected from the politico.eu website, which publishes surveys from all countries of the European Union on a regular basis, in the designated Poll of Polls section (Fig. 5). The analysis of the chart of the support for political parties from October 2021 to February 2022 shows no significant changes in the election preferences of Poles. On the other hand, continuing tendencies are visible, like the gradual drop of the support for the governing coalition (Law and Justice + Solidarity Poland) and the visible increase in the support for Civic Coalition (*Koalicja Obywatelska*) over five months.

The responses collected in the report of "Nastroje społeczne" (Social sentiments) by CBOS are divided and presented in the charts (Charts 2–4). They show the decreasing percentage of the positive responses of the supporters of Law and Justice (*Prawo i Sprawiedliwość*). Among opposition voters, no increase in the percentage of the respondents considering the situation in Poland as positive is visible.

A certain exception is Fig. 4, showing the responses of the "undecided". Apart from the visual small percentage of the voters of KO and the Left (*Lewica*), the electorate of the remaining parties seem to be more "lost" – in particular the electorate of Confederation (*Konfederacja*), which was becoming increasingly undecided in the evaluation of the situation in Poland in view of the escalating conflict at the borders of Ukraine and Russia. The electorate of Law and Justice (*Prawo i Sprawiedliwość*) shows discontent, increasing month by month and reaching more than 20 % in January. Noticeable, the increased percentage of the undecided responses was increasing every time, at the expense of the respondents that gave positive responses to the question. This is confirmed by the test conducted using the Gretl software (Fig 7). For the data from June 2021 to February 2022, negative correlation is found at –0.8727. It means that the people who began to perceive the situation as ambiguous were nearly entirely the people that had perceived the situation in Poland as positive earlier. The percentage of the undecided respondents increased in parallel with the percentage of respondents that evaluated the situation as negative. This results from the positive correlation of two variables: *negatywne_pis* and *nie_wiem_pis*, at the level of 0.5935. In addition, the results suggest that the pool of respondents with negative opinion increased practically in entirety at the expense of the group of people with positive opinion – negative correlation at –0.9109.

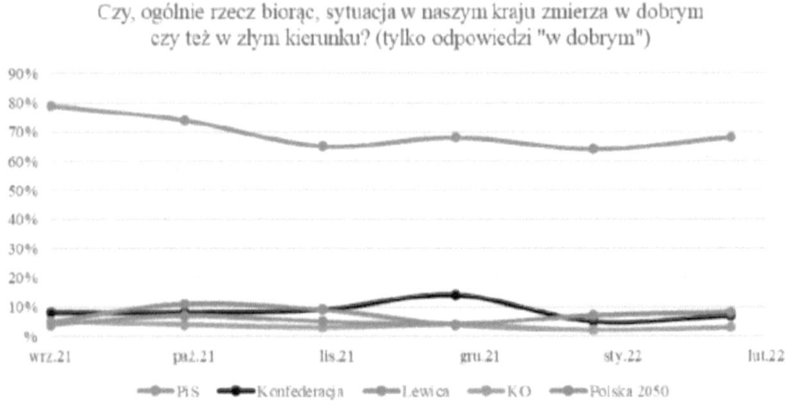

Chart 2. Social sentiments in Poland – positive responses Oct '21–Feb '22
(*Source:* original research)

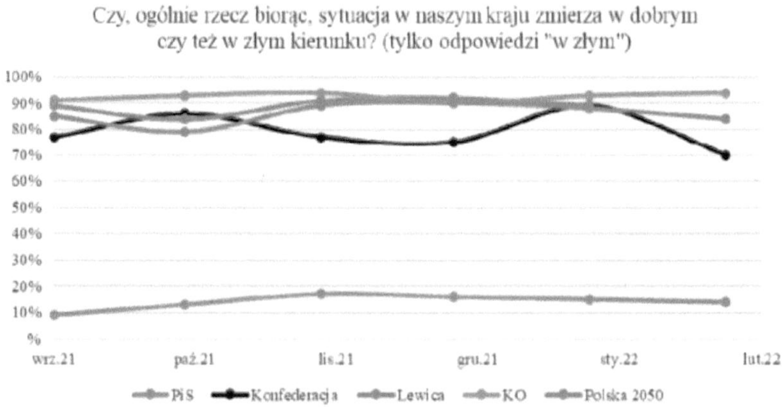

Chart 3. Social sentiments in Poland – negative responses Oct '21–Feb '22
(*Source:* original research)

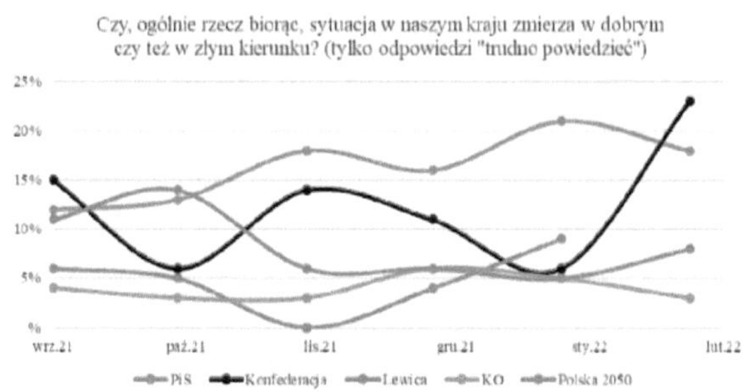

Chart 4. Social sentiments in Poland – undecided responses Oct '21–Feb '22
(*Source:* original research)

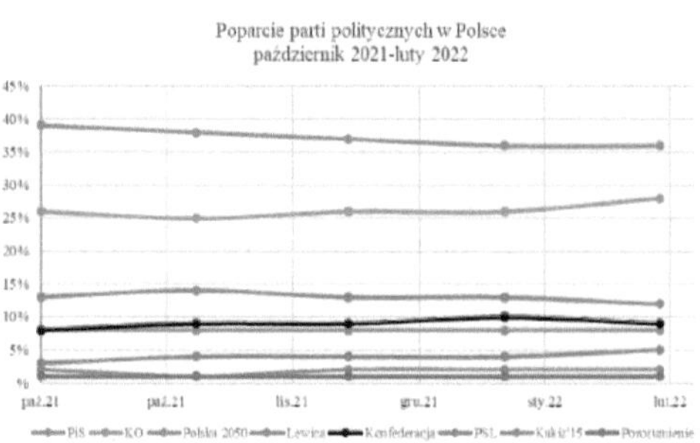

Chart 5. Support for political parties in Poland Oct '21–Feb '22
(*Source:* original research)

```
Współczynniki korelacji, wykorzystane obserwacje 1 - 9
Wartość krytyczna (przy dwustronnym 5% obszarze krytycznym) = 0,6664 dla n = 9

   poparcie_pis  pozytywne__pis  negatywne__pis  nie__wiem__pis
         1,0000          0,0283          0,1398         -0,2205 poparcie_pis
                         1,0000         -0,9109         -0,8727 pozytywne__pis
                                         1,0000          0,5935 negatywne__pis
                                                         1,0000 nie__wiem__pis
```

Fig. 1. Study of the correlation of individual response of PiS supporters in the survey "Nastroje Społeczne" (Social sentiments) by CBOS in the period

(*Source:* own study)

Despite the visible clear drop of the support for Law and Justice (*Prawo i Sprawiedliwość*) from October 2021 to January 2022 (Chart 5) and the increase in the number of respondents undecided about how to evaluate the current situation in the country (Chart 4), it cannot be determined to what extent the two variables are correlated. For this small sample, the correlation of –0.2205 cannot be treated as statistically significant, because the critical value (with two-sided critical area) for the assumed $\alpha=0.05$ and $n=9$ is 0.664. On that basis, it can only be guessed that probably as the doubts increase (along with the negative evaluation of the current situation in Poland) among the Law and Justice (*Prawo i Sprawiedliwość*) voters, the support for this party drops. This may be caused by the current scandals and by the deteriorating economic situation.

The second cyclical survey of CBOS, which was used in the study, was the survey of trust for politicians in Poland, "Zaufanie do polityków" (Trust in politicians). Contrary to the first survey, the list of politicians, in whom the level of trust may be evaluated, varies. In the CBOS report, the survey implies that the list of politicians evaluated always includes politicians like the President of Poland, the current Speakers of the respective Houses of Parliament (*Sejm* and *Senat*), the Prime Minister and the leaders of the main political parties, while the rest of the list varies. The respondents evaluated the individual representatives of the political scene, using the 11-point scale, on which the extreme points are described on one side as deep distrust (–5) and on the other side as very high trust (+5).

Based on the data collected for the period from October 2021 to February 2022, a table was generated for the net trust for the selected politicians (Tab. 3). Net trust is the difference between the percentage of respondents who trust the given politicians and those who do not trust them. Based on the data presented below, it may be clearly concluded that the trust for all politicians in the government or the governing party or connected with the governing coalition decreased in the studied period. The most important politicians of the governing coalition recorded the greatest net drops: Zbigniew Ziobro (–8), Mateusz Morawiecki (–9), and Jarosław Kaczyński (–13). The only politician of that group, for whom the net trust remained at the same level as in October, was the Minister of National Defence, Mariusz Błaszczak. At the same time, all politicians of the opposition (except for Krzysztof Bosak of Confederation (*Konfederacja*)) recorded an increase of the net trust over the studied five months: the greatest increase was recorded by Donald Tusk (+10), Rafał Trzaskowski (+9), Tomasz Grodzki and Szymon Hołownia (+7). In February, the last of the mentioned politicians, the leader of Poland 2050 (*Polska 2050*), became the politician with the highest net trust, overtaking even the incumbent President Andrzej Duda.

The juxtaposition of the survey of support for political parties and the survey "Zaufanie do polityków" (Trust in politicians), clearly show a simultaneous drop in the net trust of society in the politicians of the governing party and a drop in support for Law and Justice (*Prawo i Sprawiedliwość*) in the studied period. In turn, the politicians of Civic Coalition (*Koalicja Obywatelska*) recorded the greatest changes in the ratio of citizens with negative and positive attitude towards them. At the same time, Civic Coalition (*Koalicja Obywatelska*) gained approx. 3 percentage points in surveys.

Tab. 3. Net trust in politicians in October 2021–February 2022

	październik	listopad	grudzień	styczeń	luty	netto za okres
Andrzej Duda	14	9	6	7	13	-1
Szymon Hołownia	10	8	17	13	17	7
Mateusz Morawiecki	3	-6	-9	-7	-6	-9
Rafał Trzaskowski	-10	-4	2	-2	-1	9
Władysław Kosiniak-Kamysz	5	7	8	8	6	1
Jarosław Kaczyński	-14	-27	-26	-28	-27	-13
Zbigniew Ziobro	-12	-21	-22	-23	-20	-8
Adam Niedzielski	4	-3	-5	0	-1	-5
Donald Tusk	-27	-27	-18	-23	-17	10
Krzysztof Bosak	-5	-8	-12	-11	-8	-3
Tomasz Grodzki	-15	-10	-10	-14	-8	7
Elżbieta Witek	-1	-9	-8	-6	-7	-6
Paweł Kukiz	-15	-19	-	-	-18	-3
Piotr Gliński	1	-5	-4	-4	-1	-2
Borys Budka	-27	-	-	-	-	-
Marian Banaś	-	-7	-	-	-	-
Michał Dworczyk	-2	1	-7	-10	-8	-6
Mariusz Kamiński	-9	-10	-10	-11	-9	0
Jarosław Gowin	-19	-	-	-	-	-
Włodzimierz Czarzasty	-18	-13	-8	-18	-14	4
Przemysław Czarnek	-	-	-	-	-25	-
Jacek Sasin	-18	-26	-27	-27	-23	-5
Mariusz Błaszczak	4	0	-2	-3	0	-4
Henryk Kowalczyk	-	-	-2	-3	3	-

(*Source:* original research)

Results of the Analysis of Sentiments

The script developed in R and the sentiment dictionary modified for research purposes enabled comprehensive analysis of the collected media texts. They were investigated for the contained sentiments, divided by media outlet, interlocutor, political party and week or month of publication of the text.

To better understand the results, the plot was drawn (Chart 6) with superimposed vertical lines, presenting the most important events in Polish and European politics. These events often became important points of reference in the ongoing public debate and narrative of politicians. For particular emphasis, the events related to Pegasus surveillance were described in frames with a red outline, because of all the events that took place in October 2021–February 2022, it was the Pegasus scandal that was the greatest and most shocking.

Such presentation of data enables linking the political narrative of individual parties to specific events. The dashed lines it shows, being a graphic presentation of the most important events of the examined period, quite often coinciding with the particular "local extremes" of the weekly sentiment value of the studied parties. These are the points that could potentially result in shifts in the narrative presented by each of the described political parties. Some of them are not surprising, but they prove that all actors of the Polish political scene more or less adjust the emotional message they create to the given situation. Of course, this message differs according to the party.

One of the important culmination points was the end of the second and beginning of the third week of the studied period, when the state-owned media released two subsequent interviews with Robert Bąkiewicz, the far-right political activist from nationalist circles. His extremely negative statement in week three is also the last statement he gave in any of the described media outlets. Considering that fact and that the largest Polish media outlets select interlocutors for interviews individually, the decision to select the nationalist activist could be considered as instrumental use of his voice in the narrative on the CJEU and Constitutional Tribunal verdicts by state-owned TV.

An equally salient event that influenced the narrative of politicians was the incident at the Polish-Belarusian border, in which a group of refugees attacked the Polish border crossing with stones (week seven). The change in the narrative is visible in all parties except for PiS, which compared to the rest maintained their existing narrative at a constant level. The purpose was most probably to reassure public opinion that the governing party was in control of the situation. This crisis (in case of escalation) was potentially a very high publicity risk for the governing coalition, which may explain the strategy adopted by Law and Justice (*Prawo i*

Sprawiedliwość). Furthermore, it is no surprise that the narrative of Confederation (*Konfederacja*) changed from relatively neutral (close to 0) to extremely negative, because this party had been showing an "anti-immigrant" attitude for years.

Another important breakthrough point was the scandal of Łukasz Mejza that erupted in the ninth week of the given period. A significant increase in the weekly average sentiment value may be seen for Law and Justice (*Prawo i Sprawiedliwość*) politicians, whose clearly positive narrative (above 0) was maintained until the re-introduction of lex-TVN by PiS. This may be interpreted as an attempt to defend the politicians in the media and then undertake a media and political "offensive" to force others to change the topic of discussion to a more controversial one.

After the eruption of the scandal of Mejza, the remaining parties changed their narrative in the entirely opposite direction, shifting to a more negative narrative than before, as in the case of PSL and Agreement (*Porozumienie*). The exception was the New Left (*Nowa Lewica*), which at that time gave a series of very positive interviews, without a strong connection with current politics. In turn, after the President's veto of the media act was announced, the narrative of the opposition parties changed significantly and became more positive than before – this may be exemplified by the significant change of the weekly average sentiment value for Poland 2050 (*Polska 2050*) at the end of the thirteenth and early fourteenth week.

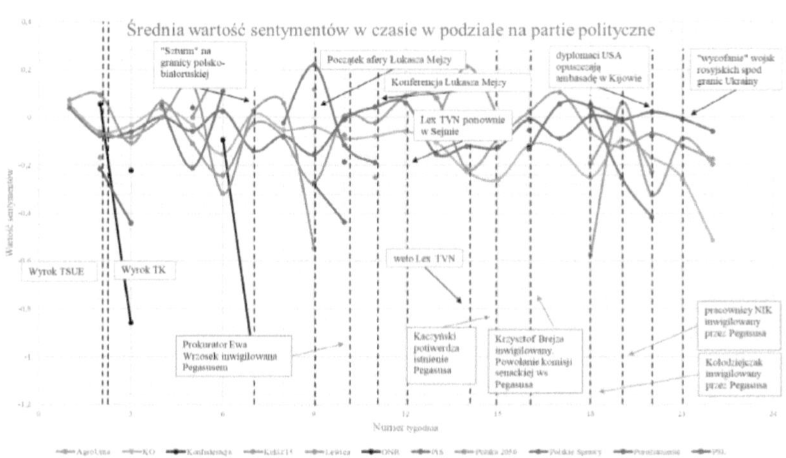

Chart 6. Average sentiment values in individual weeks in October 2021–February 2022 divided by political party with the most important events imposed

(*Source:* original research)

The only event, to which the reaction of all political circles was similar, was the escalating conflict at the Ukrainian-Russian border, which was, as we know, the direct prelude to the full-scale armed conflict. In the last week of the period under discussion, all statements of politicians of all political parties had the most negative meaning.

In addition, to complete the image, the data presented in the monthly period (Chart 7), which shows the dependencies that are difficult to capture in the more detailed weekly periods. For this reason, the smallest parties are characterised by the highest variety in their narrative over several months. It may be clearly seen that the narrative of the politicians of Jarosław Gowin's Agreement (*Porozumienie*) and Kukiz'15 changed over time. Major changes are also seen in Confederation (*Konfederacja*) and Michał Kołodziejczak, the leader of AgroUnia, which suggest the existence of a certain feature characteristic of larger parties, which consists in maintaining an apparently more constant narrative than smaller parties. The increasingly negative narrative of Civic Coalition (*Koalicja Obywatelska*) may be seen to progress month by month. A similar tendency can be seen in the politicians of PSL and Poland 2050 (*Polska 2050*).

This dependency may result from two factors. Firstly, the number of interviews given does not enable a strong impact of individual strongly negative or strongly positive statements or texts, even if they are very controversial (sometimes even scandalous) or incredibly optimistic. Of course, this refers to monthly average values. For smaller parties, the smaller number of statements subject to analysis result in an increased sensitivity of the monthly average sentiment value to individual interviews. Secondly, the number of controversial words, formulations, phrases or sentences, may disappear in the large number of other tokens, containing the words considered as neutral by the dictionary.

The detailed analysis of the results, divided by political party, shows the increasing trend in the narrative conducted by representatives of PiS. The average sentiment value in the statements at the end of the studied period is higher than at the beginning (–0.047), moving significantly towards –0.01. The range of variation of the narrative of PiS in monthly terms proved small. This probably resulted from the simultaneous attempts to launch a narrative offensive against their political opponents and to defend their own image (e.g. in December 2021, due to the Łukasz Mejza scandal and the parallel attempt to introduce lex-TVN).

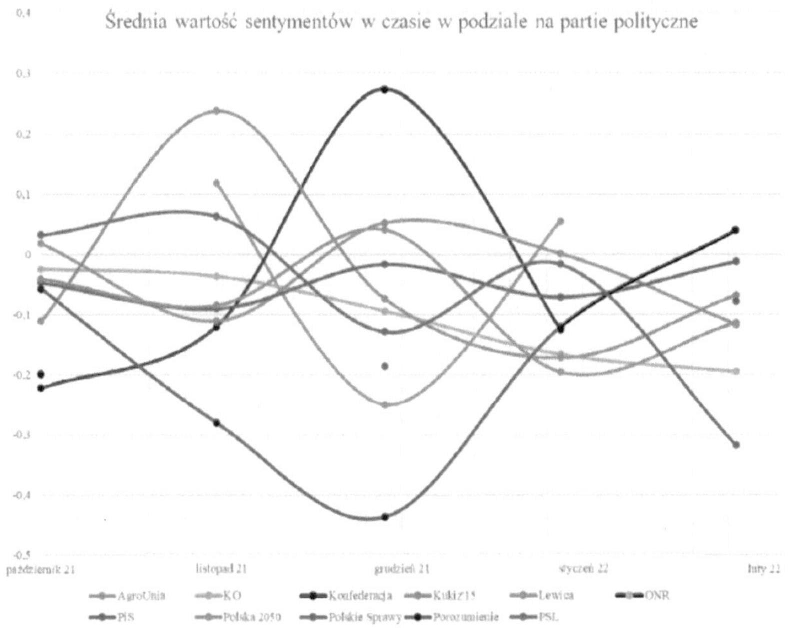

Chart 7. Monthly average sentiment values divided by political party
(*Source:* original research)

Civic Coalition (*Koalicja Obywatelska*), the second largest political group in Poland, in contrast to other actors on the political scene, was characterised by a continuously progressing drop in the average sentiment value. KO did not record even one increase of the monthly average sentiment value. In brief, the statements given by the politicians connected with KO in the media were increasingly negative every month. In October 2021, the average monthly sentiment value for KO was –0.025. At the end of the studied period, it reached the final value of –0.19. KO, in contrast to ZP, did not need to respond to a series of erupting scandals and affairs affecting the governing camp. Often the scandals present in the media space were discovered or publicised by politicians of the Civic Coalition (*Koalicji Obywatelskiej*) themselves, thus the evenly conducted narrative.

The Left (*Lewica*), in contrast to Civic Coalition (*Koalicja Obywatelska*), was characterised by a far more dynamic (variable) media narrative. The number of the collected interviews with politicians connected with the Left (*Lewica*) is only 34 % smaller than the number of interviews with the politicians of KO (205 – KO, 134 – Left). In turn, similarly to KO, the media narrative of the Left at the end of the studied period is more negative than in the first month (0.02 in October 2021 and −0.11 in February 2022). The situation is identical for Poland 2050 (*Polska 2050*), namely October 2021: −0.04, February 2022: −0.11, as well as PSL, for which the change in the narrative to more negative compared to the beginning of the studied period was even more visible, October 2021: 0.03, February 2022: −0.32.

The characteristic feature that separates the Agreement (*Porozumienie*) from the rest of the parliamentary opposition is the average sentiment value of February 2022, which is higher than that recorded at the beginning of October 2021. This difference may be explained by the more right-wing leanings of Jarosław Gowin's Agreement (*Porozumienie*) and the attempt to compete for the electorate of the governing coalition that this party had been part of only a few months before. Interestingly, the competition for the electorate of Law and Justice (*Prawo i Sprawiedliwość*), from the point of view of Agreement (*Porozumienie*) was quite uneven – the politicians of this party had been blocked from access to the state-owned and PiS-connected media for several months before the studied period [13].

Paweł Kukiz's party, Kukiz'15, who had three representatives in the Sejm of the ninth term, was considered as the coalition party of the United Right (*Zjednoczona Prawica*) (as of the time of writing of this study, the PiS Sejm club was the largest club of the Sejm of the ninth term, with 228 members of Parliament). The politicians of this party became crucial collaborators of Law and Justice (*Prawo i Sprawiedliwość*) after the break-up of the government coalition in 2021. The degree of cooperation between the two parties may be seen in the media connected with the government and Jarosław Kaczyński's party. Paweł Kukiz was practically the only important representative of a Sejm party other than Law and Justice (*Prawo i Sprawiedliwość*) or Solidarity Poland (*Solidarna Polska*), who gave an interview to Gazeta Polska. The exception was of course Marek Sawicki of PSL, however he was not the chairman of PSL. The representatives of the movement started by Paweł Kukiz were the party, of which the average sentiment value in the statements of November 2021 was the highest of all studied parties. The interview given on 6 November by Paweł Kukiz himself included many positive terms in relation to Jarosław Kaczyński, for example, he described the chairman of PiS as a warm person, who would give you the shirt off his back. However, "as politician, he would rip it off your back with flesh". In the next months, after November, the narrative presented by the party was increasingly negative, to reach its lowest level in January 2022, at the height of the Pegasus scandal, in which Paweł Kukiz advocated the appointment of the Sejm committee to investigate the illegal activity of Polish security services. In the end, the Kukiz'15 party finished the studied period with a slightly higher monthly average sentiment value than for October 2021: −0.11 compared to −0.07 in February 2022.

The study included five analysed interviews with Confederation (*Konfederacja*) politicians, none of which took place in the last studied month, February 2022. This low number of interviews, compared to the representatives of the New Left (*Nowa Lewica*), which had similar and sometimes even lower support than Confederation (*Konfederacja*), had two causes. Firstly, the politicians of this party consider social media, such as Facebook and Twitter, as their main channel of communication with their voters. The reason is the fact that a large portion of their voters or potential voters do not watch or read traditional media. This is also a consequence of the anti-mainstream character declared by Confederation (*Konfederacja*) that their voters very often identify with. The second reason, to some extent following the first, is related to the media themselves that rarely select the representatives of Confederation (*Konfederacja*) as interlocutors to interview on the most important political events, both due to their attitude towards media other than social media and the fact that the representatives of this party often use fake news, semi-truths or unscientific myths, which was particularly visible during the COVID-19 pandemic and the immigration crisis in Europe. This is the reason why interviews with Grzegorz Braun, Janusz Korwin-Mikke or Dobromir Sośnierz in the most important Polish media are rare.

However, based on the interviews that were successfully collected, the most negative narrative of the politicians of Confederation (*Konfederacja*) occurred in October 2021, in the middle of the crisis at the Polish-Belarusian border. In turn, the most positive message of that party was recorded in December of the same year, in the interview Artur Dziambor gave to Rzeczpospolita on vaccines against COVID-19. In that interview, he justified himself using the slogan "the vaccine makes you free" that appeared during a protest organised by the Confederation (*Konfederacja*). The politicians of this party try, at any cost, to present themselves as the "reasonable" part of the political scene that does not use the vaccines and scientific research that they do not consider verified. Therefore, the vocabulary and wording they use must be as appealing for their audience as possible, hence the high monthly average sentiment value.

Another set of results worth presenting is the list of the most frequent words in the given week. This list demonstrates quite clearly, what focused the political debate at the time, which in relation to the presented reference events in Chart 6 shows an integrated image of Polish politics, in particular its variability and dynamics. Furthermore, it is worth noticing the individual subjects that remained part of the infosphere created by politicians and journalists for a longer time, for example: the Pegasus surveillance or the Polish-Belarusian border.

Fig. 2. Most frequent words in the 1st, 2nd, 3rd and 4th week (left to right)
(*Source:* original research)

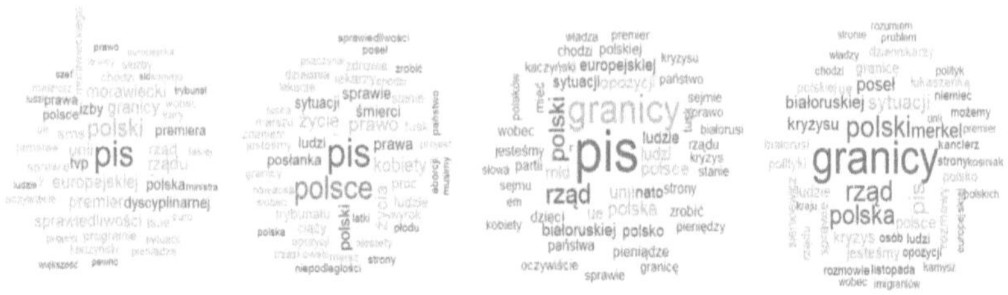

Fig. 3. Most frequent words in the 5th, 6th, 7th and 8th week (left to right)
(*Source:* original research)

The lists of the most common words are shown in Figs 2–7, shown in the form of tag clouds.

The most frequent word is "PiS". It should not be surprising, as it is the acronym of the governing party, to which every interlocutor naturally had to refer to at least once in any interview given. However, it is worth noting other words, appearing next to the word "PiS" in the individual weeks. The first studied week, or more specifically the three days from 1 to 3 October 2021, were when there was a very lively discussion on the state of emergency at the Belarusian border. In the Decree of 1 October 2011, President Andrzej Duda extended the state of emergency introduced earlier for the following 60 days [14]. At that time, the subject related to the situation at the border clearly overshadowed other important discussions on the activities of the governing party. Another hot subject was the case of what were called the children from Michałów, who in October (!) 2021 were transported by the Border Guard from the refugee centre back to the Polish-Belarusian border. Their subject only lasted in the media mainstream until the following week, which is visible in Fig. 7, which shows the tag "dzieci" (children).

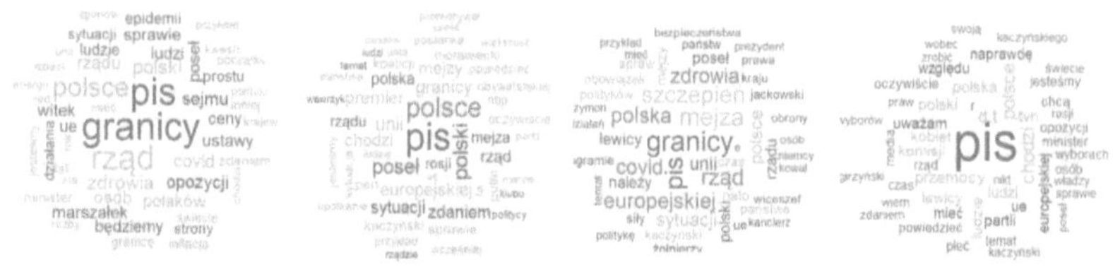

Fig. 4. Most frequent words in the 9th, 10th, 11th and 12th week (left to right)
(*Source:* original research)

Fig. 5. Most frequent words in the 13th, 14th, 15th and 16th week (left to right)
(*Source:* original research)

Fig. 6. Most frequent words in the 17th, 18th and 19th week (left to right)
(*Source:* original research)

The third and the fourth week proved to be dominated by the verdict of the Court of Justice of the European Union and the Constitutional Tribunal under Julia Przyłębska. The subject of inflation was slowly breaking into the public debate; however, as the analysis shows, that word was not frequently used. The sixth week of the study coincided with the events of Pszczyna and protests following the death of a young woman, who was a victim of the abortion regulations tightened in 2020 [15]. However, again, this subject yielded to the discussion on the situation at the Polish-Belarusian border. From the perspective of the entire studied period, one may see that regardless of the current events, the border crisis was the subject that politicians incessantly referred to in their interviews.

In the eleventh week, a temporary variation of subjects is visible. They included the scandal with Member of Parliament Łukasz Mejza in the foreground [16] or the subject of vaccinations and the COVID-19 pandemic. In addition, general security matters were addressed as well, including security at the Belarusian border. The following weeks of December included both the escalation of the Łukasz Mejza scandal, the first information on the surveillance of the politicians of the opposition and prosecutor Ewa Wrzosek with the Pegasus spy software and lex-TVN.

Fig. 7. Most frequent words in the 20th, 21st and 22nd week (left to right)
(*Source:* original research)

December was a key turning point, as emerging out of the earlier media noise was another constant component of the public debate – Pegasus. The appointment of the related Senate Committee on Pegasus and the attempt to appoint a Sejm Committee of Enquiry by votes of the opposition and Paweł Kukiz set the Polish political scene on fire. The culmination of that subject was visible in the nineteenth week of the studied period and yielded only to the events at the Ukrainian-Russian border.

The appearance of the subject of surveillance with spy software by Polish security services coincided with the tendency of the politicians of the opposition party to speak increasingly negative. This may be monitored with the calculated average sentiment from the interviews of that period (the exception is the slight increase of the average sentiment value for Poland 2050 (*Polska 2050*)).

Regardless of political divisions, the last two weeks (21 and 22) were dominated by the subject of the Russian invasion of Ukraine, which absorbed the attention of the politicians to a varied extent. Chart 8 clearly shows the reduction of the average value of sentiments in the statements of the representatives of all political parties. The greatest drop was recorded by the politicians of KO, in particular Paweł Kowal (expert on Eastern matters) and the former Minister of Foreign Affairs Radosław Sikorski, while the impact of these two interviews was clearly visible in the mentioned plot.

Conclusion

The results of this study did not show an unambiguous existence of a dependency between the political narrative and changes in the public opinion, expressed by the level of support for the individual parties in percentage points. Due to the multitude of hidden and often random factors, this result could have been fully expected and should not be surprising. However, some dependency may be seen in the correlation of the support for the individual political groupings and the evaluation of the current political situation in Poland, which is surveyed on a regular basis by CBOS. It shows that the vast majority of the electorate of the governing party have a particularly positive evaluation of the current situation in Poland (over 70 %), compared to the voters of other parties, for whom this factor reaches the values of ca. 10 %. The significant disproportion in the evaluation of the current situation points to the existence of the information bubble created by the politicians and media connected with Law and Justice (*Prawo i Sprawiedliwość*). It is impossible to indicate a specific point over the studied months, exactly at which or from which the support for Law and Justice (*Prawo i Sprawiedliwość*) started to drop. The same conclusion should be also applied to the results of the CBOS survey, in which the visible drop of positive answers is characterised by an even decrease. It may be guessed that the tactics of maintaining a relatively constant narrative in monthly terms assumed by PiS, as shown in Chart 7, resulted in only slight losses of the electorate. Nonetheless, the politicians of PiS suffered personal publicity losses, which showed in the second survey "Zaufanie do polityków" (Trust in politicians) by CBOS.

A completely different set of tactics was taken up by the opposition parties that, contrary to the governing coalition, showed a progressive shift to a more negative narrative than at the beginning of the studied period. At the same time, in most cases (apart from Civic Coalition (*Koalicja Obywatelska*)), the opposition parties were characterised by more dynamic shifts in their political narrative. In general, the governing camp presented a continually negative narrative, of which any stronger deviations only occurred in exceptional circumstances that were dangerous to the image of the governing party (e.g. the Łukasz Mejza scandal and then the introduction of lex-TVN to the Parliament to change the subject of the public debate).

Important observations are the most common words (tags) of the individual weeks. Juxtaposing them side by side enables observing the dynamics of the changing subjects. The results may suggest conclusions that are not positive. Even such important social subjects as the transportation of the children of Michałów to the forest may disappear from the course of the political narrative within one week. The dynamics of the events and the emergence of many controversial subjects in a short time resulted in quick replacement of existing subject with new subjects. This resulted in the over-saturation of the public debate with information on new scandals and controversies. The exceptions were the subjects whose presence was perceivable for several weeks in a row. However, for the most part they were directly connected with politicians in person – the Łukasz Mejza scandal or the Pegasus surveillance.

An individual evaluation of each of the discussed political parties in the context of these events in late 2021 and early 2022 enables noticing a series of differences in the strategies used to communicate with the voters and citizens. The strikingly different approach to the interpretation of the terms of propaganda, rhetoric and fake news among the politicians often leads to the distortion of the image of the present situation in the country in the eyes of a common citizen. A moral evaluation of the activities of some political parties is always entirely subjective and should be left for each individual to consider for themselves. However, some behaviours and mechanisms that are plainly in violation of the standards of the democratic rule of law and respect of the freedom of press should always be loudly and clearly condemned.

The results presented make it possible to partially organise the relatively chaotic image of Polish politics and examining the series of the subsequent events in our national infosphere.

The analyses helped to clarify the image of the independence of Polish media in late 2021 and early 2022, which unfortunately does inspire a sense of optimism. The largest Polish national media are not characterised by a sufficient level of pluralism in the selection of interlocutors, as they invite mainly representatives of the two largest political camps – the United Right (*Zjednoczona Prawica*) and Civic Coalition (*Koalicja Obywatelska*). In this summary, the worst are the state-owned media (TVP Info) and the weekly Gazeta Polska – connected to the politicians of Law and Justice (*Prawo i Sprawiedliwość*). In case of the private newspaper Gazeta Polska, this state of affairs is not surprising, but considering the high level of favour for the governing coalition in public media, it should be considered as extremely shocking and condemnable.

The computer science tools in use, enabling the analysis of sentiment in several hundred selected texts, constitute the basis for the description and identification of the contemporary sociological mechanisms that shift dynamically. Many problems presented in this study require more detailed investigation, both on the micro and macro scale. The numerous minor events combine into larger sequences of events that may last for several months or even years. As the analysed period is narrowed down, this perspective may be lost. On the other hand, some events require a more precise and careful step-by-step analysis, which is particularly important in case of different scandals related with the world of politics.

References

[1] Andrzej Piotr Stefańczyk, "Teoretyczne Założenia „Retoryki" Arystotelesa", Philosophical Review 1952, s. 368–376.

[2] Kolczyński Mariusz, "Strategie Komunikowania Politycznego", Wydawnictwo Uniwersytetu Śląskiego, Katowice 2008, ISBN 978-83-226-1656-7.

[3] Anthony Pratkanis, Eliot Aronson, "Wiek propagandy", Wydawnictwo Naukowe PWN Warszawa 2003, ISBN 83-01-140634.

[4] Elżbieta Pieniążek-Niemczuk, "On the Linguistic Features of American Political Discourse", Zeszyty naukowe Uniwersytetu Rzeszowskiego 2016 nr. 93, s. 68–77. http://doi.org/10.15584/sar.2016.13.7.

[5] "The Real Story of 'Fake News' ", Merriam-Webster.

[6] Juri Rescheto, "Wyznania Rosjanki, która była trollem. 'To się robi dla pieniędzy' ", Deutsche Welle 2018, https://p.dw.com/p/2t11A.

[7] Dominika Długosz, "'Ukrainiec-banderowiec' zastąpił 'szczepionkę, która zabija'. Propaganda ruszyła do ataku", Newsweek 2022.

[8] Cynthia Kroet, "'Post-truth' enters Oxford English Dictionary", Politico 2017.

[9] Marcin Waszak, "Postprawda i fake news czy weryfikacja treści i źródeł informacji?", Refleksje 2017 nr. 16, s. 173–188. http://doi.org/10.14746/r.2017.2.13.

[10] Nicholas O. O'Shaughnessy, Nicholas J. O'Shaughnessy, Stephan C. M. Henneberg, Stephan Henneberg, "The Idea of Political Marketing", Greenwood Publishing Group, Westport 2002, ISBN 0-275-97595-9.

[11] Robert Wiszniowski, Agnieszka Kasińska-Metryka, „Marketing Polityczny w Gorsecie Mediów", Wydawnictwo Adam Marszałek, Toruń 2014, ISBN 978-83-778-0923-5.

[12] Marta Burza, "Węgry. Ostatnia niezależna stacja radiowa traci koncesję. Krytykowała rząd Orbana. 'Wentyl prawdy' ", Gazeta.pl 2021.

[13] Kamil Dziubka, "Politycy Porozumienia odwoływani z programów TVP", Onet.pl 2021.

[14] Rozporządzenie Prezydenta Rzeczypospolitej Polskiej z dnia 1 października 2021 r. w sprawie przedłużenia stanu wyjątkowego wprowadzonego na obszarze części województwa podlaskiego oraz części województwa lubelskiego Dz.U. 2021 poz. 1788.

[15] "Ani jednej więcej protesty po śmierci Izabeli z Pszczyny aktualizowany", Oko Press 2021.

[16] "Afera Łukasza Mejzy. Wszystkie kontrowersje wokół zdymisjonowanego wiceministra sportu", Onet.pl 2021.

STATISTA

S. O'Dea, "Global internet user growth 2018–2023".

S. O'Dea, "Global number of internet users 2005–2021".

PWN

https://encyklopedia.pwn.pl/haslo/propaganda;3962718.html.

CBOS

https://www.cbos.pl/SPISKOM.POL/2021/K_162_21.PDF.

KOMUNIKAT z BADAŃ CBOS, Nr 131/2021, s. 1.

Magdalena Krutul

Emotions on News Websites During Global Crises: Analysis of Sentiment in Journalistic Publications

Introduction

In recent years, the world has been affected by significant crises. Firstly, the COVID-19 pandemic, officially announced by the WHO on 11 March 2020, which progressed on a broad scale starting in early 2020. Secondly, the outbreak of the war in Ukraine on 24 February 2022. We learned about the crises above all from the media – the popularity of Internet news websites, social media and TV stations increased significantly, both during the outbreak of the pandemic [7] and in the first weeks of the war [31]. At those times, editorial teams worked 24/7 to report the events minute by minute.

The media play a particular role during crises because they are focused not only on conveying information but also (or even above all) on conveying emotions. In the view of Werner Wirth and Holger Schramm, the media have an immense impact on the daily life of their audiences and the emotions conveyed by the media do not differ from those that we experience in person [25]. According to these researchers, apart from providing entertainment, the media may also affect behaviours and attitudes: they may set the mood by inducing the feeling of excitement, uncertainty, empathy or fear and influence our engagement and inspire us to act, which significantly increases the role of the media when there is a crisis [1].

The study presented in this article was an attempt to respond the question if the media convey more negative emotions to the audience in times of crisis. The hypothesis assumed was that during global crises, the quantity of negatively marked content in the media increases. Through the analysis of sentiment, the emotions present in the media message in the month of the outbreak of the COVID-19 pandemic (March 2020) and the month following the outbreak of the war in Ukraine (March 2022) were examined. The study material consisted of 70,492 online articles published in March 2019, 2020, 2021 and 2022 on the leading Polish news websites. The dictionary method of sentiment analysis was applied using the CLARIN-PL Literary Exploration Machine.

Sentiment Analysis: Basic Definitions and Utilities

Sentiment analysis (also: opinion mining [2]) is a method of text analysis that in an automatic or semi-automatic manner classifies the emotional markedness of statements. The purpose of sentiment analysis is to identify and classify the entire statement or its fragments for the included words with an emotional overtone [23]. This method considers two perspectives, that of the sender and the receiver, which is to say it classifies both the words that express the emotions of the sender and the words that cause emotions in the recipient [17]. Its primary application consists in reading human opinions, feelings, assessments, attitudes and emotions expressed in written language in relation to products, services, organisations, persons, matters, events, subjects and their attributes [13]. For that reason, sentiment analysis is broadly applied and used in finance, business and economy, the social and economic sciences [24], as well as, for example, in the analysis of cultural phenomena, linguistics, history and anthropology [23].

Three basic methods of sentiment analysis are distinguished – methods based on machine learning (statistical), methods based on lexicons (dictionary) and the hybrid approach [2]. Detailed classifications of the first two methods were presented by Walaa Medhat, Ahmed Hassan and Hoda Korashy [15].

The methods based on machine learning may be divided into supervised and non-supervised learning. The first requires assigning the statement to a predefined category [23], so classified text sets should be used (most often by an expert in the given field) as, for example, negative and positive. Next, machine learning algorithms are used, for example, neural networks or vector machines, to find the dependency between the individual statement features and the learning set. This method is most often used for analysis at the document level [5]. Non-supervised learning, however, is applied based on hierarchical methods, methods based on similarity and methods based on text complexity measures [23]. This approach checks the occurrence of specific words, the syntax of the statement, the punctuation and the emoticons used [24].

The methods based on lexicons use the definitions of words [2] included in dictionaries created manually, automatically or semi-automatically [24]. Dictionaries enable classification based on:

- key expressions, for example, *key word in context* (KWIC analysis), *key phrase in context* (KPIC analysis), subject analysis;
- the logical conditions of the text content, for example, single- and multi-argument operators, contextual operators;
- lexical rules, for example, similarities and differences in word meanings, language grammar, and language word resource [23].

The combination of statistical and dictionary methods is referred to as the hybrid approach. It consists in using the *quickness of the dictionary approach and the high precision of machine learning effects* [2]. In the view of some researchers, this approach is the most effective in sentiment analysis [27].

Sentiment Analysis as the Method of Investigation of Texts in Polish

Some of the methods presented above may be used regardless of the language of the analysed texts. Other models, however, require training based on the same language or application of appropriate lexicons. In this study, sentiment analysis was applied to articles in Polish, therefore it seems appropriate to quote information on the existing lexicons required for the sentiment analysis in Polish.

Probably the most popular set is the Nencki Affective Word List (NAWL [41]), which includes 2,902 Polish words – popular nouns, verbs and adjectives – annotated with the type of emotion evoked. These words were characterised as evoking or associated with the following basic emotions: happiness, anger, sadness, fear and disgust. The database was developed by the researchers of the Nencki Institute in 2015, based on *the Berlin Affective Word List Reloaded (BAWL-R)* [19], the list of affective words in German. The study resulting in the annotation of words involved 266 people. The purpose was to create a standardised database as point of reference for, for example, psychological, cognitive and linguistic research [28].

Another available set is the sentiment dictionary created by the Team of Linguistic Engineering at the Department of Artificial Intelligence at the Institute of Computer Science of the Polish Academy of Sciences (IPI PAN). The lexicon consists of 1,200 adjectives and adverbs that were manually marked and verified by two linguists and then expanded by 2,600 words algorithms and machine learning [53]. The sentiment in this dictionary is presented on several scales:

- two-grade (positive/negative)
- three-grade (positive/neutral/negative)
- five-grade (very negative/negative/neutral/positive/very positive)
- nine-grade (using a paradigm based on singular value distribution (svd) [26]).

The largest set including words with specific markedness is Wordnet. It is a semantic dictionary that includes 191,000 words, including as many as 80,000 words with an emotional annotation [37]. The dictionary items in Wordnet include much information on the relationships between the word units (synonymy, antonymy, derivativeness etc.), including extended information on the markedness, that is: polarisation, emotions, valuation and examples of use.

Wordnet was created through a collaboration between lexicographers and computer scientists of the Group of Language Technologies of the Wrocław University of Technology under the European CLARIN (Common Language Resources & Technology Infrastructure) scientific infrastructure, created to facilitate working with large text sets to researchers in humanities [38]) and the Polish CLARIN-PL scientific consortium. This is a significant resource used for the computer processing of the Polish language and research on artificial intelligence [37]. The works on the Polish Wordnet started in 2005 [6]. Wordnet is available for public (also commercial) use [40].

Applications for Sentiment Analysis

Sentiment analysis can be conducted using the analytical tools available on the Internet, which are universally accessible tools within the CLARIN scientific infrastructure.

To determine the sentiment of shorter text fragments in Polish, SentEMO [44] can be used. This is an open network system that enables entering the text in the text field or uploading a corpus to determine the polarisation of the emotional overtones in three visualisations – dot plots, histograms and pie charts – with the following information:

- no sentiment
- positive polarity (strong)
- positive polarity (light, weak)
- negative polarity (strong)
- negative polarity (light, weak)
- ambiguous overtones [12].

The platform is available at: https://ws.clarin-pl.eu/sentemo.shtml.

Another tool made available by CLARIN is Sentiment. This tool is used to verify the markedness by the basic types of emotion and the universal values assigned to words. The results are presented in three plots. The first determines the general emotional markedness by presenting the number of occurrences of the words described as positive or negative. The second presents basic emotions: gladness, trust, enjoying something expected, sadness, anger, fear, disgust, and surprise with something unexpected. The third visualises universal values: utility, happiness, uselessness, harm, error and unhappiness [9]. The tool is available at: https://ws.clarin-pl.eu/sentyment.

Another easily accessible tool is MultiEmo – the application for multi-language sentiment analysis designed to examine eleven languages, including

Polish [48]. The demo version enables text analyses at three different levels: text, sentence and paragraph. In addition, the tool provides automatic language detection. The analysis results are provided in the form of distribution of the following parameters: *plus*, *amb.*, *zero* and *min*. The sentiment analysis is presented in tables and in bar charts, line charts and pie charts, which allows visualisation of the distribution of the emotional overtones, the values of the decision function and the distribution of the occurrence of the given tag in the text [43]. The tool is available at: https://ws.clarin-pl.eu/multiemo.

A particularly well-developed tool that easily processes larger data sets is the Literary Exploration Machine (LEM) [14]. This is a system that processes texts and extracts statistical information, while offering integrated tools:

- Apache Tika [42] – the file converter, detects and extracts text from many different file types (e.g.: doc, docx, pptx, xlsx, odt, pdf, html, rtf).
- Morfeusz 2 with the SGJP dictionary [34] – performs morphological analysis.
- WCRFT2 [36] (…) – the morpho-syntactic tagger.
- Liner2 (…) – the tool used, for example, to determine proper names.
- WSD [46] (…) – the disambiguation of lexical meanings.
- Wordnet – the semantic dictionary used to disambiguate meanings and to find hyponyms and hyperonyms.
- WebSty [45] – used to determine the similarity and group texts.
- NKJP [35] – the used tagset (system of tags) [29].

One of the functionalities of the tool is to perform sentiment analysis. In addition, the utility enables loading and using a custom list of stopwords, that is, words to exclude from the analysis (usually words that do not contribute meaningful content to the message: conjunctions, pronouns, prepositions etc.) [47]. Another advantage of the application is that the Literary Exploration Machine accepts input data in the form of text corpora up to 15 MiB with many extensions uploaded in ZIP files, via URL address, the DSpace repository [39] or from the CLARIN-PL Nextcloud cloud drive. The output data, however, is obtained, depending on the task selected, in xlsx, csv, txt or archive format. LEM also integrates the functions of the SentEMO and Sentiment tools discussed earlier, although the presentation of data differs. Instead of graphical visualisations, the sentiment analysis in LEM is presented in a table (xlsx) with the following example categories:

- number of tokens
- number of verbs and pronouns in the 1st, 2nd and 3rd person of singular and plural

- number of words with neutral, positive and negative polarisation
- number of words with assigned emotions: uselessness, error, sadness, anger, disgust, harm, utility, trust, gladness, happiness, unhappiness, truth, fear, knowledge, good, surprise with something unexpected, enjoying something expected, beauty, ugliness, ignorance.

The tool is available at: https://ws.clarin-pl.eu/lem.

Commercial Solutions and Tools

The sentiment analysis of texts in Polish can also be performed using commercial solutions. Services of this kind are provided by companies such as Sentimenti, SentiOne and Brand24. These companies rely on proprietary algorithms using, for example, Internet big data corpora, CLARIN-PL solutions, artificial intelligence, machine learning, neural networks and research with human involvement. The information on the models used is not publicly available.

Media during Crisis

During crises, the media play an important role, because they bring the audience closer to the event, influence their emotions and define the threat. The media message may augment, reduce and downplay the risk felt [1]. The presentation of events is so important that it affects the engagement of the audience. Many publications and reports were developed to describe the phenomena that occurred in the media during the crises of recent years. This chapter describes some of them.

Media during the Pandemic

During the pandemic, our daily lives were reorganised [4]. At that time, many people experienced over-saturation with media contact and indicated the necessity of being selective with them. At the same time, interpersonal communication took place on the Internet on a larger scale [22]. In the first weeks of the pandemic, ca. 25 % of workers in Poland worked remotely, while internet traffic increased by ca. 30 % [7]. The pandemic to some extent forced contact with the media, due to frequently changing regulations on the rules for being in public spaces. That was the time of the record increase in the traffic on websites compared to the year before, which is shown in Chart 1.

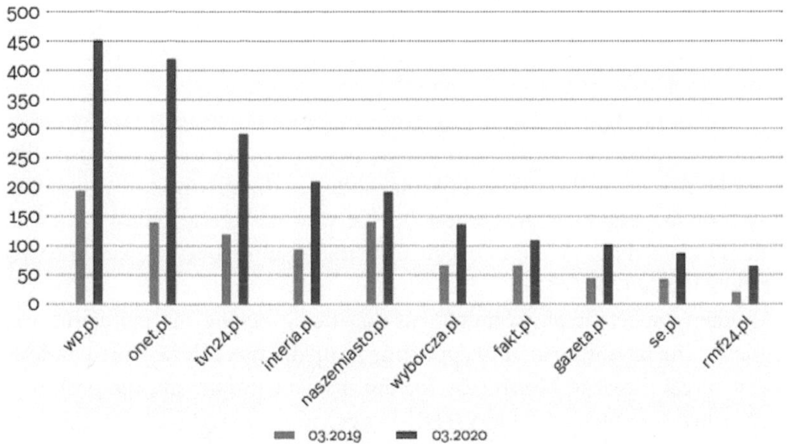

Chart 1. Number of impressions of the most popular Polish news websites in March 2019 and March 2020 (in M)

Source: J. Grzeszak, "Modele biznesowe mediów po pandemii", Polski Instytut Ekonomiczny, Warszawa 2021, p. 9.

As a result of the outbreak of the pandemic, trends observed for many years in the media accelerated, i.e. drop in the sales of print press and increased traffic on websites. Research conducted in December 2020 showed that as many as 44 % of people used websites more for information than before the pandemic [7].

At the same time, despite their wide reach, they struggled with many problems, for example, financial (one of the reasons was the reduction of spending on advertisement by many companies), personal (redundancies caused by cost reductions), restricted access to sources (limits on the number of journalists at press conferences, concerns among physicians about talking to journalists), the necessity to adjust to the technical requirements of remote work (e.g. radio broadcasts from home, remote interviews) [18].

The media adjusted to the new situation by producing bulletins and guides related to health and economics on a broader scale [18]. Media formats such as newsletters and podcasts were appreciated [7]. The homepages of news websites began to show information on the daily numbers of people infected with the coronavirus, the number of tests for SARS-COV-2 completed and later the number of persons vaccinated. The role of experts in the media increased as well. Academics in the social sciences and humanities were more eagerly invited to conversations on the future, cultural trends, social changes and transformations of reality. The experts in these fields began to shape public opinion [3].

The media informed of the scale of the hazards, while indicating how people should behave in face of the hazards (communication on the distancing, disinfection, mask principle or calls to stay at home). From a sociological perspective, the media is seen as the source of social changes, the driving force to act and the role of shaping social awareness. In the view of Teresa Sasińska-Klas, the pandemic caused the media to become the unquestioned first power in society on a global scale [20].

Another important phenomenon in the media during the pandemic were *fake news*. The false information appearing in media space during the pandemic became a real threat to health (e.g. the anti-vaccine movement, the promotion of unverified treatments for COVID-19) or economic security (the triggering of panic and irrational consumer behaviours, for example, mass cash payouts after the false news that the government planned on requisitioning the private savings of citizens)[16].

Media during the War in Ukraine

The fact that the media change during times of crisis could also be observed at the time of the Russian invasion of Ukraine. Then the media, regardless of the subjects they addressed on a daily basis, were dominated by information on the outbreak of the war for several weeks. After the Russian invasion of Ukraine, news websites recorded an increase in their reach of more than 30 % in the first week of the war [30]. The interest was slowly decreasing – from 24 February to 2 March, the news websites in the *News and journalism–general* category were visited by 12.3M users per day, while in the second week of war the average number of users was 10.7M, to drop to 10M in the third week [49].

The media are of great importance during an armed conflict, because – as proved many times in research – the way in which events are presented has a direct impact on the "knowledge, attitudes, emotions, behaviours of the society, as well as the reputation of the people presented in the media" [10]. The format of all-day coverage was particularly engaging and gave the feeling of involvement in the events. At that time, almost every new piece of information was presented as sensational and the format transformed into some kind of infotainment [10].

Polish media took a clear stand of opposition against the aggression of Russia on an independent state and sided with Ukraine. The support was expressed by, for example, numerous fundraising initiatives for refugees, creation of content in Ukrainian, sharing information on ways to help or the symbolic change of logos to Ukrainian colours. The attitude of engagement in the media caused a series of phenomena. First, the manner of reporting of the war events was favourable for the Ukrainian side, which influenced the visual and rhetorical creation [11]. Second, the power of the media to inspire action and building engagement triggered a huge movement of opposition in the audience against politicians connected with Russia or acting in its interest. In particular, on news websites and in social media calls were made to boycott the products of the companies that did not withdraw from Russia [52]. Another expression of support and a stimulus to trigger engagement among the people was the appeals in the media, previously unpublished photos of army columns and transports of weapons and military equipment [32]. It was shown that the media message has the potential to convey the actual experience of war [8]. Thus, it may be claimed that the components engaging to act and the strongly emotional messaging had an impact on the audience's feeling of involvement in the resistance movement. In this case, "common" Internet users, through their Internet activity (at different levels: from activism up to cyberterrorism) found themselves at the front in cyberspace [21].

Another symptom of the online war was the propagation of false information. In the context of the Russian invasion of Ukraine, the information war is discussed as the new way to wage war against the enemy. Such operations were of particular importance in Russia, where the media justify the aggression and create an alternative reality in the context of the reasons for the military operations. In turn, the *fake news* propagated beyond the borders of Russia was intended to exert psychological pressure on selected societies [51]. The scale and effectiveness of these operations are reflected in the comments of disinformation experts, who claim that even major press agencies happened to replicate unverified information [50].

Analysis of Emotions on News Websites during Crises

The study used the corpus of 70,492 articles sourced from seven websites: onet.pl, wp.pl, interia.pl, gazeta.pl, o2.pl, wyborcza.pl, fakt.pl. In recent years, these websites topped the rankings in terms of reach and opinion-forming. The study involved all publications of March 2019, 2020, 2021 and 2022, in different subject categories (e.g. the country, world, Poland, business, information, events, weather, gossip, health, sports, and horoscopes), mostly marked with the umbrella category of *news* or *events* or *information*. The number of publications from each period and website are presented in Tab. 1.

Tab. 1. Number of publications in the research material in analysed periods

Website	March 2019	March 2020	March 2021	March 2022
Fakt	3,126	2,659	3,589	3,566
Gazeta.pl	1,248	1,456	1,938	2,016
Interia	2,265	2,855	2,245	2,847
o2	503	1,261	2,610	2,580
Onet	4,234	3,778	2,788	3,231
Wirtualna Polska	2,228	2,039	3,082	2,678
Wyborcza.pl	2,822	2,445	2,072	2,331

(*Source:* original research)

To highlight the different subjects dominating in the studied period, a word frequency study was conducted. The data is presented in the following word clouds (Figs 1, 2, 3 and 4).

Fig. 1. Word cloud from articles of March 2019

Fig. 2. Word cloud from articles of March 2020

Fig. 3. Word cloud from articles of March 2021

Fig. 4. Word cloud from all studied articles of March 2022

Sentiment Polarisation Analysis

Sentiment polarisation was studied from two perspectives. First, in holistic terms, the share of words with positive and negative sentiment was examined for the entire research material of the given year. Second, in terms of the dominant positive or negative sentiment per article was considered and the percentage share of articles with dominant positive and negative sentiment was analysed. The results are shown in Plots 2 and 3.

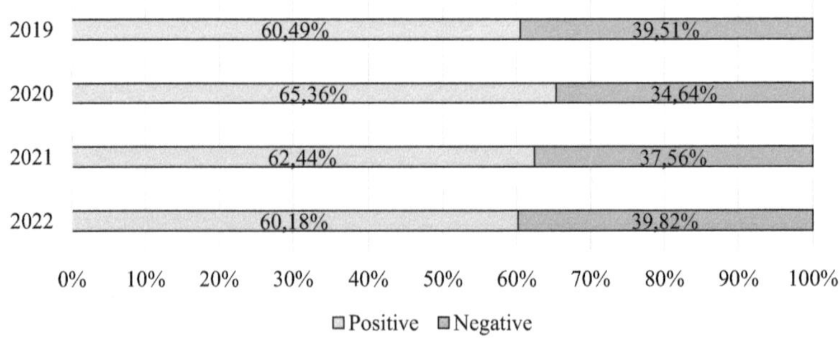

Plot 2. Share of positive and negative sentiment in research material in March of each year

(*Source:* own study)

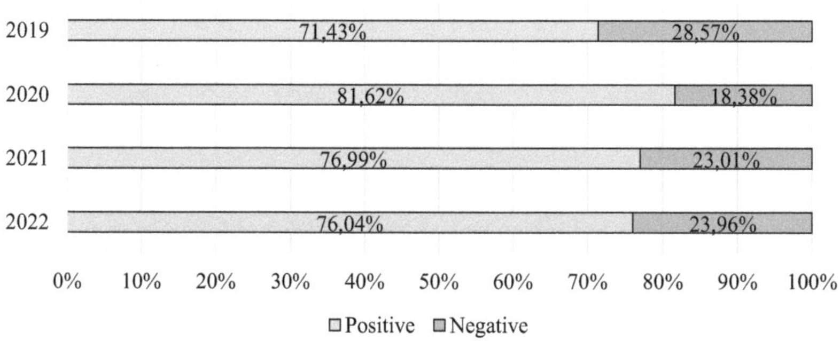

Plot 3. Share of articles with dominant positive or negative sentiment in March of each year
(*Source:* own study)

The polarisation – from both perspectives – was also analysed with specification of values for each website. The data showing the share of positively and negatively marked content for the entire material and for the dominant polarity per article are shown in Tabs 2 and 3, respectively.

Tab. 2. Share of positive (+) or negative (–) polarity in holistic terms for studied websites in March of each year

Website\Year of publication	**2019**		**2020**		**2021**		**2022**	
	+	–	+	–	+	–	+	–
Fakt	59.9 %	40.1 %	64.3 %	35.7 %	63.2 %	36.8 %	60.7 %	39.3 %
Gazeta.pl	58.2 %	41.8 %	66.5 %	33.5 %	63.4 %	36.6 %	59.0 %	41.0 %
Interia	61.6 %	38.4 %	66.7 %	33.3 %	63.6 %	36.4 %	60.6 %	39.4 %
o2	55.0 %	45.0 %	61.1 %	38.9 %	62.2 %	37.8 %	61.7 %	38.3 %
Onet	61.6 %	38.4 %	67.6 %	32.4 %	61.9 %	38.1 %	62.6 %	37.4 %
Wirtualna Polska	58.7 %	41.3 %	67.3 %	32.7 %	64.7 %	35.3 %	57.4 %	42.6 %
Wyborcza.pl	61.3 %	38.7 %	61.8 %	38.2 %	58.7 %	41.3 %	58.2 %	41.8 %

(*Source:* original research)

Tab. 3. Share of articles predominantly positively (+) or negatively (–) marked for the studied websites in March of each year

Website\Year of publication	**2019**		**2020**		**2021**		**2022**	
	+	–	+	–	+	–	+	–
Fakt	70.6 %	29.4 %	78.8 %	21.2 %	77.5 %	22.5 %	73.0 %	27.0 %
Gazeta.pl	66.1 %	33.9 %	84.8 %	15.2 %	78.6 %	21.4 %	75.6 %	24.4 %
Interia	70.9 %	29.1 %	81.7 %	18.3 %	77.8 %	22.2 %	76.4 %	23.6 %
o2	55.5 %	44.5 %	72.1 %	27.9 %	74.8 %	25.2 %	72.6 %	27.4 %
Onet	70.4 %	29.6 %	81.8 %	18.2 %	74.4 %	25.6 %	81.6 %	18.4 %
Wirtualna Polska	65.0 %	35.0 %	81.8 %	18.2 %	75.2 %	24.8 %	66.8 %	33.2 %
Wyborcza.pl	84.3 %	15.7 %	87.0 %	13.0 %	82.7 %	17.3 %	87.0 %	13.0 %

(*Source:* original research)

The analysis in holistic terms (Tab. 2) shows the highest share of positive polarity – 67.6 % – in March 2020 at onet.pl and the highest share of negative polarity in March 2019 at o2.pl. However, based on the data presented in Tab. 3. it may be stated that the highest share of articles dominantly positive markedness in all studied periods was found at Wyborcza.pl.

To clearly illustrate the changes in the share of articles predominantly negatively marked, year to year, Tab. 4 presents the difference in the share of sentiment compared to the previous year in percentage points.

Tab. 4. Change (difference) in the share of articles predominantly negatively marked for the studied websites in March each year, year to year comparison

Website\Year of publication	**2020**	**2021**	**2022**
Fakt	–8.20 %	1.30 %	4.50 %
Gazeta.pl	–18.70 %	6.20 %	3.00 %
Interia	–10.80 %	3.90 %	1.40 %
O2	–16.60 %	–2.70 %	2.20 %
Onet	–11.40 %	7.40 %	–7.20 %
Wirtualna Polska	–16.80 %	6.60 %	8.40 %
Wyborcza.pl	–2.70 %	4.30 %	–4.30 %

(*Source:* original research)

On each of the analysed websites, in 2020 the share of articles with predominantly negative sentiment decreased. In 2021, on six of seven websites, the share of articles with predominantly negative sentiment increased. In turn, in the analysed year 2022, an increase in articles with predominantly negative sentiment was found on five of seven of the analysed websites, year to year. To answer the question as to what the percentage change of the share of articles predominantly negatively marked was, the data were also compared year to year also from this perspective. The results are shown in Tab. 5.

Tab. 5. Percentage decrease or increase in the share of articles predominantly negatively marked for the studied websites in March each year, year-to-year comparison

Website\Year of publication	**2020**	**2021**	**2022**
Fakt	–27.89 %	6.13 %	20.00 %
Gazeta.pl	–55.16 %	40.79 %	14.02 %
Interia	–37.11 %	21.31 %	6.31 %
o2	–37.30 %	-9.68 %	8.73 %
Onet	–38.51 %	40.66 %	-28.13 %
Wirtualna Polska	–48.00 %	36.26 %	33.87 %
Wyborcza.pl	–17.20 %	33.08 %	-24.86 %

(*Source:* original research)

The highest percentage decrease in the share of articles predominantly negatively marked, at 55.16 %, was recorded at Gazeta.pl in 2020, following the outbreak of the pandemic. In the following year 2021, the share of articles with dominant negative markedness was found to increase above 30 % at Gazeta.pl, Onet, Wirtualna Polska and Wyborcza.pl. At the time of the outbreak of war, the highest increase in articles predominantly negatively marked, at 33.87 %, was recorded on the Wirtualna Polska website.

Positive, Negative and Neutral Sentiment

The sentiment analysis in the LEM utility also enables analysis of data divided into positive, negative and neutral sentiments. Tabs 6 and 7 present the study results from a holistic perspective. The highest values of each sentiment are in bold.

Tab. 6. Share of neutral, positive and negative sentiment in publications of March of each year

Sentiment\Year of publication	2019	2020	2021	2022
Neutral	76.34 %	**77.42 %**	76.54 %	74.79 %
Positive	14.31 %	14.76 %	14.65 %	**15.17 %**
Negative	9.35 %	7.82 %	8.81 %	**10.04 %**

(*Source:* original research)

Tab. 7. Share of neutral (0), positive (+) and negative (–) sentiment for studied websites in March of each year, as percentage (%)

Website\ Year of publication	2019			2020			2021			2022		
	0	+	–	0	+	–	0	+	–	0	+	–
Fakt	74.2	15.5	10.4	76.1	15.4	8.5	74.5	16.1	9.4	73.8	15.9	10.3
Gazeta.pl	77.8	12.9	9.3	79.1	13.9	7.0	78.5	13.6	7.9	76.3	14.0	9.7
Interia	79.2	12.8	8.0	**79.5**	13.6	6.8	79.1	13.3	7.6	78.5	13.0	8.5
O2	73.3	14.7	12.0	75.1	15.2	9.7	73.9	16.3	9.9	73.6	16.3	10.1
Onet	77.6	13.8	8.6	79.2	14.1	6.7	79.0	13.0	8.0	75.4	15.4	9.2
Wirtualna Polska	77.4	13.3	9.3	77.7	15.0	7.3	78.7	13.8	7.5	75.7	13.9	10.4
Wyborcza.pl	73.1	16.5	10.4	73.0	16.7	10.3	71.7	16.6	11.7	70.8	**17.0**	**12.2**

(*Source:* original research)

Based on the data presented above, the share of contents with positive and negative overtones was the highest in March 2022. At that time, the vocabulary described as neutral had the lowest share compared to other analysed periods. March 2022 also marked the highest share of words with negative and positive sentiment for the research material at Wyborcza.pl. The highest share of contents neutrally marked was recorded in the research material from Interia, at 79.5 % in March 2020.

Analysis of Eight Basic Emotions

At the next stage of research, the analysis was conducted according to Plutchik's wheel of emotions [33], divided into eight basic emotions. The share of individual emotions in the research material is presented in Tab. 8. The highest shares for individual emotions are written in bold.

Tab. 8. Sentiment share of eight emotions in publications in March of each year

Sentiment/Year of publication	2019	2020	2021	2022
JOY	24.11 %	**25.79 %**	25.49 %	23.93 %
SADNESS	20.23 %	20.72 %	20.08 %	**21.23 %**
ANGER	**19.65 %**	17.90 %	18.82 %	18.85 %
TRUST	14.54 %	**15.84 %**	14.99 %	15.80 %
FEAR	6.49 %	6.58 %	6.17 %	**6.94 %**
DISGUST	**8.03 %**	6.08 %	7.17 %	7.17 %
ENJOYING SOMETHING EXPECTED	4.85 %	5.00 %	**5.16 %**	4.37 %
SURPRISE WITH SOMETHING UNEXPECTED	2.09 %	2.08 %	**2.11 %**	1.72 %

(*Source:* original research)

The highest share of JOY compared to all studied periods occurred in the research material of March 2020 and the highest share of emotions SADNESS and FEAR was recorded in March 2022. In March 2020 and March 2022, higher values were also recorded for the share of TRUST, compared to other analysed periods. The detailed percentage increases and decreases of the share of emotions, year to year, are presented in Tab. 9.

Tab. 9. Percentage decrease or increase of the sentiment share of the 8 basic emotions in publications in March of each year, in year-to-year comparison

Sentiment/Year of publication	2020	2021	2022
JOY	6.97 %	−1.16 %	-6.12 %
SADNESS	2.42 %	−3.09 %	5.73 %
ANGER	−8.91 %	5.14 %	0.16 %
TRUST	8.94 %	−5.37 %	5.40 %
FEAR	1.39 %	−6.23 %	12.48 %
DISGUST	−24.28 %	17.93 %	0.00 %
ENJOYING SOMETHING EXPECTED	3.09 %	3.20 %	−15.31 %

(*continued on next page*)

Tab. 9. Continued

Sentiment/Year of publication	2020	2021	2022
SURPRISE WITH SOMETHING UNEXPECTED	−0.48 %	1.44 %	−18.48 %

(*Source:* original research)

As seen in the above results, the highest percentage drop was recorded for DISGUST, of which the share in 2020 decreased by as much as 24.28 % compared to 2019. The year 2022 recorded an 18.48 % decrease in the emotion of SURPRISE WITH SOMETHING UNEXPECTED and a 15.31 % decrease in ENJOYING SOMETHING EXPECTED. The highest percentage increases in the share are related to DISGUST in March 2021 (17.93 %) and FEAR in March 2022 (12.48 %). It is also worth noting the percentage decrease in the share of ANGER (8.91 %) and increase in TRUST (8.94 %) in March 2020.

Another proposed analytical perspective is the comparison of the data for the years 2020, 2021 and 2022 to that for 2019, before the outbreak of the coronavirus pandemic (Tab. 10).

Tab. 10. Percentage decrease or increase of the sentiment share of the eight basic emotions in March of each year, compared to the year before the pandemic (2019)

Sentiment/Year of publication	2020	2021	2022
JOY	6.97 %	5.72 %	−0.75 %
SADNESS	2.42 %	−0.74 %	4.94 %
ANGER	−8.91 %	−4.22 %	−4.07 %
TRUST	8.94 %	3.09 %	8.67 %
FEAR	1.39 %	−4.93 %	6.93 %
DISGUST	−24.28 %	−10.71 %	−10.71 %
ENJOYING SOMETHING EXPECTED	3.09 %	6.39 %	−9.90 %
SURPRISE WITH SOMETHING UNEXPECTED	−0.48 %	0.96 %	−17.70 %

(*Source:* original research)

Considering the changes exceeding 4.9 % compared to 2019, the following is apparent:

- in 2020 (mentioned above), increases in TRUST and JOY, decreases in DISGUST and ANGER;
- in March 2021, visible increases in ENJOYING SOMETHING EXPECTED and JOY, decrease in DISGUST and FEAR;
- in March 2022, percentage decrease in the share of SURPRISE WITH SOMETHING UNEXPECTED, DISGUST, ENJOYING SOMETHING EXPECTED, increase in TRUST, FEAR and SADNESS.

The image of the differences in the emotions presented above may also be seen by comparing the values to the average data for all studied periods, which is shown in Chart 4 and Plot 5.

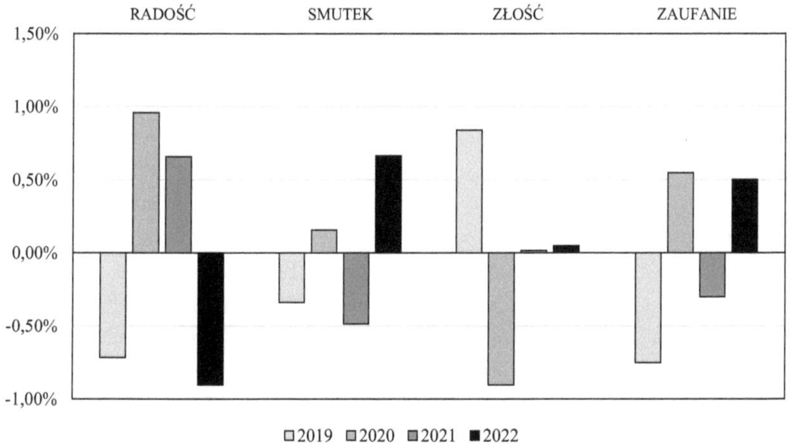

Chart 4. Change in share of emptions: JOY (Radość), SADNESS (Smutek), ANGER (Złość), TRUST (Zaufanie) in March of each year, compared to average

(*Source:* original research)

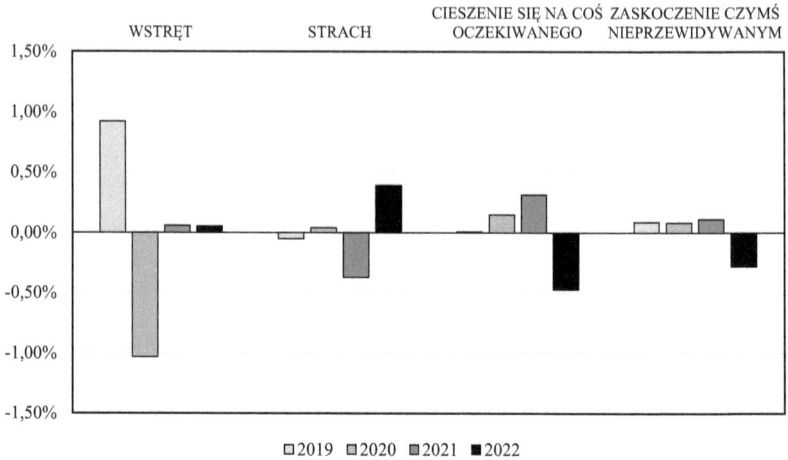

Plot 5. Change in share of emptions: DISGUST (Wstręt), FEAR (Strach), SURPRISE WITH SOMETHING UNEXPECTED (Cieszenie się na coś oczekiwanego), ENJOYING SOMETHING EXPECTED (Zaskoczenie czymś nieprzewidywanym) in March in each year, compared to average

(*Source:* original research)

Under the assumed criterion of difference above 0.5 % compared to the average, March 2020 saw a higher level of JOY and TRUST, as well as lower level of ANGER and DISGUST. However, March 2022 recorded a higher share of SADNESS and TRUST, as well as lower level of JOY. The reference periods, however, were characterised by in March 2019, a higher level of ANGER and DISGUST, as well as lower level of JOY and TRUST; in March 2021, an above-average share of JOY.

Sentiment Analysis of Articles with Selected Keywords

The above analysis may be complemented with an analysis of the articles on the crises themselves, extracted from the research material. For this purpose, a list of keywords related to the pandemic crisis in March 2020 and the outbreak of war in Ukraine in March 2022 was created. The following keywords were selected for the research material of 2020: coronavirus, COVID-19, quarantine, epidemic, and pandemic. 11,644 articles of the total of 16,493 articles of 2020 included one or more of the above keywords (70.6 %). The keywords assumed to extract the articles related to the outbreak of war were war, Ukraine, Russian, Putin, and Ukrainian. The number of articles with these keywords was 13,733 of the total of 19,249 (71.3 %). This high share of the keywords in the analysed material confirms that during the crisis, the media are focused above all on the related subjects. Some articles refer to the crises directly, some of them contain text references and links to materials on the crisis.

The results of the sentiment analysis for this group of articles are presented with full data for the materials of March 2020 and March 2022. To facilitate the presentation of data, the tables use the terms *pandemic* and *war* to refer to the research material including the keywords related to each subject.

The following Chart 6 shows that the negative polarisation for the research material with keywords related to the pandemic has a 1.25 percentage point lower share than all articles of that period. In articles related to war, however, the increase of the negative polarity of 1.28 percentage point is found in comparison with all articles of the analysed month.

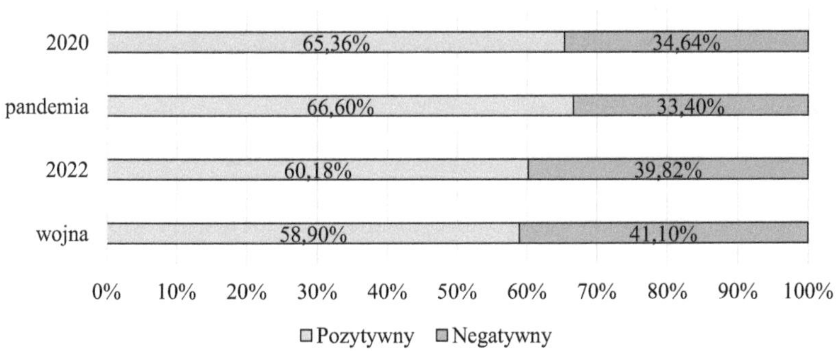

Chart 6. Sentiment polarity in research material of March 2020 and March 2022, as well as articles with selected keywords – holistic perspective
(*Source:* original research)

The second perspective, determining the share of articles predominantly positively and negatively marked, is presented in Chart 7.

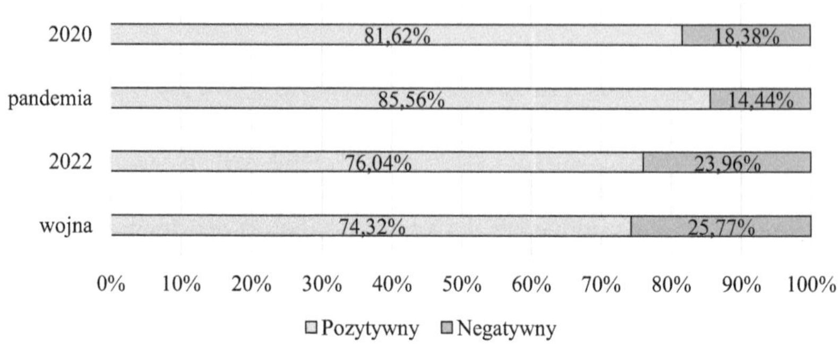

Chart 7. Share of articles with predominantly positive or negative sentiment in March 2020 and March 2022, as well as articles with selected keywords
(*Source:* original research)

In the analysis of the predominantly positive or negative marking per article, the data also indicate the decrease in the share of the negative sentiment for the articles related to the pandemic and the increase for the articles related to war.

The comparison of data regarding neutral, positive and negative sentiments (Tab. 11) shows the highest share of the negative sentiment for the group of articles with keywords related to war (10.41 %) and the highest share of positive sentiment for all articles of March 2022 (15.17 %). The highest share of words with neutral markedness was recorded for the group of articles with keywords related to the pandemic (78.07 %).

Tab. 11. Share of neutral, positive and negative sentiment in publications of March 2020 and March 2022, as well as articles with selected keywords

Sentiment\Group of publications	2019	2020	Pandemic	2021	2022	war
Neutral	76.34 %	77.42 %	**78.07 %**	76.54 %	74.79 %	74.66 %
Positive	14.31 %	14.76 %	14.60 %	14.65 %	**15.17 %**	14.92 %
Negative	9.35 %	7.82 %	7.32 %	8.81 %	10.04 %	**10.41 %**

(*Source:* original research)

The analysis of emotions in articles with keywords related to the pandemic and war (Tab. 12) shows some tendencies highlighted for the full material of the respective months. In the group of publications with keywords related to the pandemic, the highest shares were recorded for JOY and TRUST. In the research material with keywords related to the war in Ukraine, the highest shares of SADNESS and FEAR may be indicated, compared to other analysed groups of publications.

Tab. 12. Sentiment share of eight emotions in publications of March 2020 and March 2022, as well as articles with selected keywords

Sentiment/Group of publications	2019	2020	Pandemic	2021	2022	war
JOY	24.11 %	25.79 %	**25.81 %**	25.49 %	23.93 %	22.70 %
SADNESS	20.23 %	20.72 %	21.08 %	20.08 %	21.23 %	**21.73 %**
ANGER	**19.65 %**	17.90 %	17.69 %	18.82 %	18.85 %	19.23 %
TRUST	14.54 %	15.84 %	**16.29 %**	14.99 %	15.80 %	16.10 %
FEAR	6.49 %	6.58 %	6.67 %	6.17 %	6.94 %	**7.17 %**
DISGUST	**8.03 %**	6.08 %	5.52 %	7.17 %	7.17 %	7.32 %
ENJOYING SOMETHING EXPECTED	4.85 %	5.00 %	4.89 %	**5.16 %**	4.37 %	4.12 %
SURPRISE WITH SOMETHING UNEXPECTED	**2.09 %**	2.08 %	2.05 %	2.11 %	1.72 %	1.63 %

(*Source:* original research)

In this case, it is in particular worth noting the differences in emotions between the group of articles with keywords related to the pandemic and war. The contents related to the pandemic have a 3 percentage point higher share of JOY and 1.8 percentage point lower share of DISGUST compared to the content on the war.

Summary and Conclusions

This study made it possible to observe in the publications of March 2020 a significant presence of subjects related to the coronavirus pandemic. However, this proved to have had no impact on the increase in the share of negative sentiment on news websites, compared to March 2019. On the contrary, in the month of announcement of the global pandemic the share of positive sentiment was the highest in comparison with other analysed periods. The greatest decreases in the share of articles predominantly negatively marked compared to 2019 were recorded at Gazeta.pl (55.16 %) and the smallest decreases at Wyborcza.pl (17.2 %).

The emotions on the websites at the time of the outbreak of the pandemic slightly deviated from the distribution of emotions in the subsequent years. What may be seen in the analysed periods is the highest share of JOY and, compared to March 2020 and 2021, the increases of SANDESS and TRUST and the decreases OF DISGUST and ANGER. According to Plutchik's wheel, joy and trust are positive emotions, while sadness, disgust and anger are negative emotions [33].

Complementing the study with an analysis of articles with keywords related to the coronavirus pandemic confirmed the phenomena observed in the full material of that year. The articles with keywords related to the pandemic also showed higher values of the share of positive sentiment, the highest of all groups of articles analysed in this study. In addition, the data related to the polarisation and share of articles with dominant positive and negative sentiment, as well as the data on eight emotions, confirmed the assumptions that the subjects related to the crisis are what contribute to the increase of positive sentiment and decrease of negative sentiment. Nearly all tendencies observed in relation to the full material of 2020 may be confirmed with more confidence for the material with keywords. Only in the distribution of emotions, it is possible to notice the higher percentage share of ENJOYING SOMETHING EXPECTED for the full material of 2020 than for the group with keywords.

The increase in the positive sentiment of the content in 2020 is demonstrated by all analysed perspectives. Thus, it may be said that at the outbreak of that crisis, the share of content with negative sentiment dropped.

2021 was also part of the period, in which the pandemic took place. The word cloud shows, however, that the subjects related to the coronavirus were not as popular as at the outbreak of the pandemic. The word cloud for publications of 2021 is more similar to the one produced based on the material of 2019 than in 2020. The articles published one year after the outbreak of the pandemic show higher negative sentiments than the ones of March 2020. At the same time, the share of negatively marked content in 2021 is still lower than in 2019. The emotions on news websites in 2021, compared to the period before the pandemic, were characterised by the increase in JOY and ENJOYING SOMETHING EXPECTED, as well as the decrease in DISGUST, FEAR and ANGER.

The articles written in March 2022, directly after the outbreak of the war in Ukraine, are characterised by a higher share of negative sentiment compared to all previous analysed periods and by the increase in article with dominant negative sentiment compared to 2020 and 2021. Also the highest share of content with positive and negative sentiments was recorded, as well as the lowest share of neutral sentiment of all analysed periods (in the comparison of shares of neutral, positive and negative sentiment), which may evidence the more frequent use of words with emotional overtones by the journalists.

In addition, the increases in SADNESS and FEAR that reached the highest shares of the analysed periods do not seem to be insignificant. The increase in the percentage share of FEAR, compared to 2021, was 12.48 %. The study also recorded an increase in TRUST compared to 2021 and the highest share of this emotion in relation to other groups of materials.

At this point, it is worth noting that in publications of March 2022, the dominant word was *Ukraine* and in the research material as much as 71.3 % of articles were publications with keywords related to the outbreak of war. The study on that group of articles showed an even greater share of negative sentiment in the article content, smaller share of neutral content and higher share of negative content (in the division into neutral, positive, negative) and the lower level of JOY, higher level of ANGER and FEAR than for the entire material of 2022.

The war in Ukraine and the coronavirus pandemic, even though both phenomena may be referred to as recent crises, did not cause similar emotion in the media. It is worth noting that the group of articles with keywords related to the pandemic included a higher share of neutral words, while the group of publications with keywords related to war had the smallest share of neutral words of all the analysed publications. The common feature, as it may seem characteristic of the occurrence of crisis, is the increase in the share of the emotions of TRUST and FEAR.

Conclusion

The study was dedicated to the analysis of the emotional layer of journalistic publications published for the top news websites in Poland during world crises. The hypothesis of this study was that during crises the quantity of contents with negative overtones in the media increases. The study provided the grounds to reject the hypothesis thus formulated.

Based on the sentiment analysis, in the period of the outbreak of the pandemic, news websites released relatively more words with positive overtones and more articles with dominant positive markedness. In the period of the beginning of the war in Ukraine, however, the tendencies were the contrary. The websites released more negatively marked content and more articles with predominantly negative sentiment compared to the analogical periods. In view of the above, the hypothesis cannot be confirmed, because its veracity is only confirmed by one of the two analysed cases.

A possible extension of the study may be the analysis of articles written in the analogical time a month before, In February 2019, 2020, 2021 and 2022, to check if compared to the relevant periods and escalation of crises, changes are visible in the markedness of the texts.

References

[1] Augustowska Karolina, Pietkiewicz Jana, "Kryzysy w mediach. o emocjach, zaangażowaniu i epatowaniu cierpieniem" [w:] "Obywatelskość 2.0: zaangażowane narracje o obywatelskości młodych mieszkańców Litwy i Polski", red. M. Jakimowicz i M. Sikora, Oficyna Wydawnicza Epigram, Bydgoszcz 2016.

[2] Baj-Rogowska Anna, "Analiza sentymentu jako narzędzie monitorowania wyników finansowych przedsiębiorstwa", Przedsiębiorczość i Zarządzanie t. XX, z 12, cz. 1, 2019, s. 125–139.

[3] Bogunia-Borowska Małgorzata, Sarzyńska Anna, "Pandemia koronawirusa. Analiza dyskursu naukowego i eksperckiego w polskich mediach" [w:] "Komunikowanie w procesie zmian", red. W. Świerczyńska-Głownia, A. Hess, M. Nowina Konopka, Kraków 2021, s. 105–108.

[4] Drozdowski Rafał i in, "Życie codzienne w czasach pandemii. Raport z pierwszego etapu badań", Wydział Socjologii UAM, Poznań 2020.

[5] Dzieciątko Mariusz, Spinczyk Dominik, "Text mining. Metody, narzędzia, zastosowania", Wydawnictwo: Wydawnictwo Naukowe PWN, Warszawa 2016, ISBN: 9788301185107.

[6] Dziob Agnieszka, Łazarewicz Paulina, "Słowosieć jako narzędzie wspomagające pracę tłumacza", Rocznik Kognitywistyczny V, (2011), https://doi.org/10.4467/20843895RK.12.004.0408.

[7] Grzeszak Jacek, „Modele biznesowe mediów po pandemii", Polski Instytut Ekonomiczny, Warszawa 2021, ISBN 978-83-66698-34-5.

[8] Guittet Emmanuel-Pierre, Zevnik Andreja, "Exposed images of war" [w:] "Emotions, Politics and War", red. L. Ĺhäll, T. Gregory, Londyn 2015, s. 192.

[9] Janz Arkadiusz, Kocoń Jan, Piasecki Maciej i Zaśko-Zielińska Monika, "plWordNet as a Basis for Large Emotive Lexicons of Polish", Conference: LTC'17 8th Language & Technology Conference Poznań 2017.

[10] Klepka Rafał, "Wojna w mediach: wybrane zagadnienia dotyczące relacjonowania konfliktów zbrojnych". Wojna i konflikty: Przeszłość – Teraźniejszość – Przyszłość, nr 1 (1) (2016).

[11] Kampka Agnieszka (por.), "Obraz – retoryka – wartości. Medialne relacje wojenne". Roczniki Nauk Społecznych, t. 14 (50), nr 2, (2022), s. 21–39. http://doi.org/10.18290/rns22502.2.

[12] Kocoń Jan, Miłkowski Piotr, Zaśko-Zielińska Monika, "Multi-Level Sentiment Analysis of PolEmo 2.0: Extended Corpus of Multi-Domain Consumer Reviews", Proceedings of the 23rd Conference on Computational Natural Language Learning (CoNLL) 2019, http://dx.doi.org/10.18653/v1/K19-1092.

[13] Liu Bing, "Sentiment Analysis and Opinion Mining", Morgan & Claypool Publishers, May 2012, https://www.cs.uic.edu/~liub/FBS/SentimentAnalysis-and-OpinionMining.pdf.

[14] Maryl Maciej, Piasecki Mariusz, Walkowiak Tomasz, "Literary Exploration Machine A Web-Based Application for Textual Scholars" [w:] "Selected papers from the CLARIN Annual Conference 2017", Budapest, 18–20 September 2017, Linköping University Electronic Press nr. 147, (2018) s. 128–144.

[15] Medhat Walaa, Hassan Ahmed, Korashy Hoda, "Sentiment analysis algorithms and applications". A survey, Ain Shams Engineering Journal t. 5, (2014).

[16] Mroczka Kamil, „Fake newsy jako nowa kategoria zagrożenia systemu bezpieczeństwa ekonomicznego państwa w dobie kryzysu epidemicznego". Przegląd Bezpieczeństwa Wewnętrznego, w. 14, nr 26 (2022), s. 86–121.

[17] Pang Bo, Lee Liliian, "Opinion Mining and Sentiment Analysis", Foundations and Trends in Information Retrieval nr 2 (2008), s. 1–135. http://dx.doi.org/10.1561/1500000011.

[18] Popielec Dominika, "News deserts, digital transformation and remote work. Media in the face of problems and challenges during the pandemic in the bimonthly 'Press'" [w:] "Komunikowanie w procesie zmian", red. W. Świerczyńska-Głownia, A. Hess, M. Nowina Konopka, Kraków 2021, s. 129–142.

[19] Rigiel Monika i inni, "Nencki Affective Word List (NAWL): the cultural adaptation of the Berlin Affective Word List–Reloaded (BAWL-R) for Polish". Behavior Research Methods nr 47, (2015), s. 1222–1236. http://doi.org/10.3758/s13428-014-0552-1.

[20] Sasińska-Klas Teresa, "Media w czasach płynnej pandemii – perspektywa socjologiczna" [w:] "Komunikowanie w procesie zmian", red. W. Świerczyńska-Głownia, A. Hess, M. Nowina Konopka, Instytut Dziennikarstwa, Mediów i Komunikacji Społecznej Uniwersytetu Jagiellońskiego, Kraków 2021, s. 13–14. ISBN: 978-83-66492-15-8.

[21] Stoppel Anna, "Wojna w sieci, czyli drugie oblicze wojny w Ukrainie" [w:] "Edukacja dla bezpieczeństwa. Przegląd naukowo-metodyczny", red. J. Buczyński i inni, Poznań 2022.

[22] Szulich-Kałuża Justyna, Pastwa Rafał, "Wstęp" [w:] "Media w dobie pandemii COVID-19", red. R.J. Pastwa, Wydawnictwo KUL, Lublin 2021.

[23] Tomanek Krzysztof, "Analiza sentymentu – metoda analizy danych jakościowych. Przykład zastosowania oraz ewaluacja słownika RID i metody klasyfikacji Bayesa w analizie danych jakościowych". Przegląd Socjologii Jakościowej nr 10/2 (2014).

[24] Turek Tomasz, „Możliwości wykorzystania analizy sentymentu w procesach prosumenckich". Ekonomiczne Problemy Usług, t. 2, nr 1 (126), (2017), s. 286–288.

[25] Wirth Werner, Schramm Holger, "Media and Emotion". Communication Research Trends nr 24 (2005), s. 8–15.

[26] Wawer Aleksander, "Mining Co-Occurrence Matrices for SO-PMI Paradigm Word Candidates" [In:] Proceedings of the Student Research Workshop at the 13th Conference of the European Chapter of the Association for Computational Linguistics, EACL'12 SRW, pp. 74–80, Avignon, France, April 2012a. Association for Computational Linguistics.

[27] Ziora Leszek, "The sentiment analysis as a tool of business analytics in contemporary organizations", Studia Ekonomiczne. Zeszyty Naukowe Uniwersytetu Ekonomicznego w Katowicach, Katowice nr 281, 2016.

Internet Sources

[28] "Emocjonujące! Powstała lista polskich słów emocjonalnych", pozyskano z: https://naukawpolsce.pl/aktualnosci/news%2C406223%2Cemocjonujace-powstala-lista-polskich-slow-emocjonalnych.html [accessed: 03.09.2022].

[29] "LEM instrukcja użytkowania", pozyskano z: https://nextcloud.clarin-pl.eu/index.php/s/MFdA7CR5MVlypZC#pdfviewer [accessed: 03.07.2022 r.].

[30] "Serwisy informacyjne po rosyjskiej agresji na Ukrainę – analiza", pozyskano z: https://media-panel.pl/pl/aktualnosci/serwisy-informacyjne-po-rosyjskiej-agresji-na-ukraine-analiza/ [accessed: 10.09.2022 r.].

[31] "WP Raport: polski rynek mediowy w obliczu wojny w Ukrainie", Wydanie I, pozyskano z: https://reklama.wp.pl/files/dvr6ijxy63aqi3utr4zt/Raport_rynek%20mediowy%20w%20czasie%20inwazji%20na%20Ukrain%C4%99.pdf [accessed: 03.09.2022 r.].

[32] Brzeziński Maciej, "Widzisz transport wojskowy, robisz zdjęcia? Nie pokazuj i nie udostępniaj! To kwestia bezpieczeństwa", pozyskano z: https://www.auto-swiat.pl/wiadomosci/aktualnosci/widzisz-transport-wojskowy-robisz-zdjecia-nie-pokazuj-to-kwestia-bezpieczenstwa/4w91edc [accessed: 11.09.2022 r.].

[33] Czostka Agnieszka, "8 emocji podstawowych. Po co badamy ich aż tyle?", pozyskano z: https://sentimenti.pl/blog/dlaczego-mierzymy-az-8-emocji-podstawowych/ [accessed: 10.09.2022 r.].

[34] http://morfeusz.sgjp.pl/ [accessed: 03.07.2022 r.].

[35] http://nkjp.pl/poliqarp/help/ense2.html [accessed: 03.07.2022 r.].

[36] http://plwordnet.pwr.wroc.pl/wordnet/ [accessed: 03.07.2022 r.].

[37] https://clarin-pl.eu/ [accessed: 03.07.2022 r.].

[38] https://clarin-pl.eu/dspace/ [accessed: 03.07.2022 r.].

[39] https://exp.lobi.nencki.gov.pl/nawl-analysis [accessed: 03.09.2022].

[40] https://tika.apache.org/ [accessed: 03.07.2022 r.].

[41] https://ws.clarin-pl.eu/multiemo [accessed: 03.07.2022 r.].

[42] https://ws.clarin-pl.eu/sentemo.shtml [accessed: 03.07.2022 r.].

[43] https://ws.clarin-pl.eu/websty [accessed: 03.07.2022 r.].

[44] https://ws.clarin-pl.eu/wsd [accessed: 03.07.2022 r.].

[45] https://www.seopilot.pl/wiki/Stopwords.html [accessed: 03.07.2022 r.].

[46] Kocoń Jan, Miłkowski Piotr, Kanclerz Kamil, "MultiEmo: Multilingual, Multilevel, Multidomain Sentiment Analysis Corpus of Consumer Reviews" CLARIN-PL digital repository, 2021, http://hdl.handle.net/11321/798.

[47] Kozielski Maciej, "Mediapanel: zainteresowanie informacjami w sieci nadal wyższe niż przed wojną", pozyskano z: https://www.press.pl/tresc/69951,mediapanel_-zainteresowanie-informacjami-w-sieci-nadal-wyzsze-niz-przed-wojna [accessed: 10.09.2022 r.].

[48] Radkowski Michał, "Fake newsy elementem wojny na Ukrainie. Kinga Klich z Demagoga: 'nawet duże agencje prasowe się mylą' ", pozyskano z: https://www.wirtualnemedia.pl/artykul/fake-news-dezinformacja-wojna-ukraina-rosja-teorie-spiskowe-fact-checking [accessed: 11.09.2022 r.].

[49] Ryba Anna, "Działania informacyjne jako uzupełnienie wojny konwencjonalnej", pozyskano z: https://www.wojsko-polskie.pl/woc/articles/publikacje-r/dzialania-informacyjne-jako-uzupelnienie-wojny-konwencjonalnej/pdf/ [accessed: 11.09.2022 r.].

[50] Szymczak Jakub, "Lista wstydu. Zobacz, które firmy odmawiają bojkotu i nie wycofują się z Rosji", pozyskano z: https://oko.press/firmy-ktore-nie-wycofaly-sie-z-rosji/ [accessed: 11.09.2022 r.].

[51] Wawer Aleksander, "Związki między wyrażeniami wydźwięku a atrybutami opinii", pozyskano z: https://docplayer.pl/166443690-Atrybutami-opinii-aleksander-wawer-2-grudnia-2013.html [accessed: 03.08.2022 r.].

Bogusław Jagusiak and Patryk Szczepkowski

The Significance of Social Media in the Disinformation Process

Introduction

The information sphere in the 21st century has developed to such a degree that the only necessary tools are the Internet and a device connected to it. This provides every entity worldwide with access to information, both from the world and from their own country. There are exceptions, in which the citizens of the given country do not have such access, a good example being the North Korea, where all information is checked by the authorities. Information is an essential problem in the media world, because it may take various forms and consequently cause threats, both to citizens and to states. A major role in the dissemination of information is played by the media, one of the factors proving it is information control, also known as censorship. This phenomenon is used to hide some information from the public opinion, while presenting it with other, equally relevant events. The media world along with the developing technology is one of the most important challenges for humankind, because improper use of the media may lead to irreversible consequences.

Social Media

One of the most important technologies of the 21st century is social media. Broader dissemination of information, not only for the selected people, but for any user, has become possible due to social media. The media themselves are based on Web 2.0 technology. This term was first used in 1999[1]; Web 2.0 technology covers the following activities that the user may perform using social media[2]:

- publish content
- comment on content

1 D. DiNucci, *Fragmented Future*, "Print" 1999, 4 (53), p. 32.

2 G. Fleming, Global Social Technographics Update 2011: US and EU Mature, Emerging Markets Show Lots of Activity, https://www.forrester.com/blogs/12-01-04-global_social_technographics_update_2011_us_and_eu_mature_emerging_markets_show_lots_of_activity/ [05.04.2022].

- react
- share
- organise groups.

The above activities to a large extent describe the capabilities of social media. Aichner and Jacob mention thirteen types of media in their article on social media. They include, for example, blogs, different types of online boards, as well as common projects (Wikipedia) or virtual game worlds.[3]

One of the most important roles in social media is the entities publishing information. In addition, the following user roles may be distinguished[4]:

- creators
- interlocutors
- reviewers
- collectors
- visitors
- viewers
- inactive.

Each of the above groups constitutes forms that enable the users to actively participate in the discussion. Seven groups of social media users are distinguished. One of the largest groups is content creators, with the obligation to verify and disseminate true information. However, enforcing this responsibility is relatively negligible. In social media, information may be manipulated by commentators or viewers, but only within the subject addressed by the creator.

Another component to mention following the activities on the subject, which is the information, is the division of contents posted in social media. The information may take various forms, including[5]:

3 T. Aichner, F. H. Jacob, *Mearuring the Degree of Corporate Social Media Use*, Article available at: https://www.researchgate.net/publication/283073224_Measuring_the_Degree_of_Corporate_Social_Media_Use [11.12.2021 r.].

4 G. Fleming, Global Social Technographics Update 2011: US and EU Mature, Emerging Markets Show Lots of Activity, https://www.forrester.com/blogs/12-01-04-global_social_technographics_update_2011_us_and_eu_mature_emerging_markets_show_lots_of_activity/ [05.04.2022].

5 L. Floridi, Philosophical Conceptions of Information, [in:] Formal Theories of Information, Sommaruga G., Ed. Springer, 2009.

- true information – information consistent with the reality and based on facts,
- misleading information – untrue information, provided to the public opinion in a non-intentional manner,
- false information – intentionally false, misleading and alleged information, aimed at influencing the actions of the society,
- malicious information – true information, disclosed intentionally to harm a given entity,
- unidentified information – information that cannot be verified for truthfulness and intentionality due to technology or lack of knowledge.

All of the above-mentioned types of information are shared on social media. It should be noted that influencing and manipulating is visible not only in the published content, but also in the comments or reactions. The most undesirable information is false information, because it has the most negative affect on society. It can be used by certain entities to influence the communities or specific groups, which causes disturbances, divisions and, in the worst case, may have an impact on human health and lives. In addition to false information, the media may use other mechanisms to fight against undesirable and inconvenient information. They include[6]:

- censorship – restricting access to information by preventing access. Information that is detrimental to public opinion and inconvenient may remain unpublished.
- propaganda – publication and promotion of information that are beneficial from the point of view of the sharing entity.
- information over-saturation – the process intended to prevent the verification of information.

Electronic media, due to their particular traits, differ from traditional media. The main differences between the media in terms of information dissemination include[7]:

- reach
- financial cost
- informing entity
- content reproduction

6 R. Brzeski, *Dezinformacja,* Article available at: https://socjocybernetyka.files.wordpress.com/2012/02/dezinformacja-skrypt-1.pdf [15.06.2022].

7 P. Bączek, *Dezinformacja jako zagrożenie dla bezpieczeństwa informacyjnego współczesnych społeczeństw,* [in:] Security Review 2020 no. 4(17), pp. 3–18.

- dissemination rate
- content control
- responsibility.

These categories show on one hand how electronic media have unlimited reach, low financial costs, unlimited numbers of entities sharing information, unlimited content reproduction, immediate dissemination, no control and negligible responsibility. On the other hand, the traditional media, which include television, radio and press, are the opposite of this.

The end result of each piece of information disseminated are activities and behaviours of the environment. When false information/disinformation is disseminated, one speaks of the disinformation impact. Undoubtedly, the longer false information is propagated, the stronger the disinformation impact is. In psychology, the purpose of disinformation is to add to witness testimony information inconsistent with the course of the event, from a source outside of the event itself.[8] The capabilities to influence and manipulate in the dissemination of information are apparent both in social media and in traditional media. The principles of exerting influence on the society include[9]:

- the principle of reciprocity
- the principle of obligation
- the principle of resolution
- the principle of shortage
- the principle of authority
- the principle of cohesion and engagement
- the principle of sympathy
- the principle of social proof.

The principles of exerting influence on society are mechanisms that can be seen in the dissemination of information. To convince another person to take an action, the author of the content may subtly use the above elements. In the media, however, the most visible components of manipulation also have certain principles[10]:

8 R. Polczyk, M. Szpitalak, *Efekt wzmocnionej afirmacji – Wzrost odporności na dezinformację wskutek autoafirmacji wzmocnionej pozytywną informacją zwrotną*, Psychological studies tome 50 (2012), vol. 3, pp. 63–64.

9 Ch. Hadnagy, *Socjotechnika. Sztuka zdobywania władzy nad umysłem*, Ed. Onepress 2020.

10 Ibid.

- increased susceptibility
- forced evaluation
- force deprivation
- punishment
- intimidation
- control of surroundings.

As exerting influence, manipulation is a set of activities of an entity to convince others or oneself to take action or refrain from it. One of the most effective forms of manipulation is the multitude of aspects, which is when both sides obtain benefits from the activities undertaken.[11]

Social media have an advantage in the dissemination of disinformation over traditional media. As mentioned above, the capability to disseminate information in electronic media is greater.

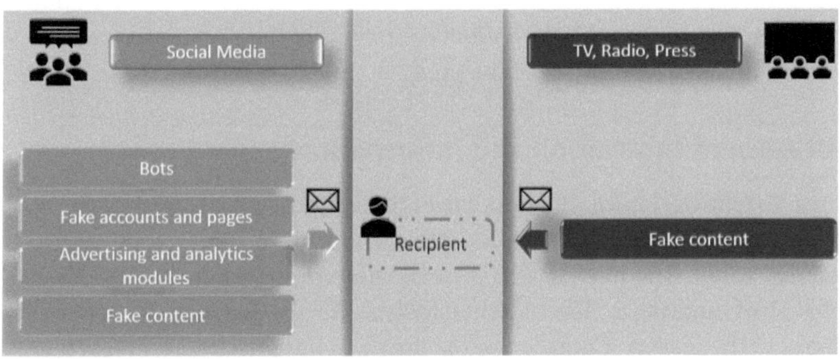

Fig. 1. Tools for the dissemination of information
Source: original work based on: https://www.youtube.com/watch?v=hLSNdEWxmv8 [05.04.2022].

The above items show the difference between the media in terms of capabilities and tools used to disseminate information. In traditional media, these are only the false contents shared, for example, in news programmes. In social media, these tools are more numerous. The media users may come from all over the worlds and still comment, provide opinions or evaluate the information. Thus, the tool referred to as

11 M. Skała, *Manipulacja odczarowana! 777 skutecznych technik wpływu*, Ed. Helion, Gliwice 2015, p. 19.

bots/trolls appeared, the purpose of which is to agitate the opinion of other users by use of extreme emotions. Such activities may be both internal and external.[12]

Another component are false accounts or pages, which may influence society, for example, by appearing to be authoritative. Due to its nature, this tool may cause much harm, from changing behaviours to the loss of funds.

The final component differing the traditional media and electronic media are the analysis and advertisement modules. This capability to analyse and collect data is a significant component, above all interest is visible in the sphere of marketing and politics. There, the examination of the behaviour of society may contribute to the purchase or winning elections. The same analysis includes one more factor, that is, the capability of selecting users, to whom information must be delivered. After analysis, the addition of the stimulus to encourage behaviour may convince the users to act as appropriate.

Undoubtedly, social media have a significant advantage over traditional media. Perhaps with time they may replace television, radio or the traditional press. A noticeable trend is the shift of operations of the traditional media to the Internet, in particular in case of the press.

Diffusion of Innovation and Information

In the virtual world, information has become a product made available to the users. The behaviours and activities prove if the given information caused the intended result. The information dissemination process may be described as a diffusion of innovation. This term involves the process that introduces new innovations to the market. Considering the theme of this article, our innovation will be information. The diffusion process includes knowledge[13]:

- psychology
- strategy formulation
- market analysis
- marketing
- technology.

Diffusion is a type of transformation in the functioning or structure of the given community. The result of the activity is rejection or acceptance of the new innovation. This process includes four components: innovation, communication

---

12 D. Jachyra, *Trollowanie – antyspołeczne zachowania w Internecie, sposoby wykrywania i obrony*, [in:] Working Papers of University of Szczecin, no. 656 (28) 2011, pp. 253–254.

13 E. M, Rogers, *Diffusion of Innovations*, Ed. Free Press, New York 2003, p. 5.

channel, time and social system. Diffusion may be perceived as spreading of the product from the source to the end community.[14] The rejection or acceptance of a new innovation by the user is determined by attributes. These include[15]:

- comparative advantage – innovation perception level
- compatibility – degree of perception a compatible with the existing values, needs and experiences,
- complexity – level of understanding or application. The simpler, the easier to spread.
- testing capability – level of capability for testing or checking innovation,
- observability – degree of notability of innovation for other users.

Among all creators of information in social media, the creation of content based on the above components provides competitive advantage or retention of information in the public opinion. The evidence of the success of the created information is its dissemination. At this point, it is worth returning to the matter of tool difference between the social media and traditional media. This tool are the analysis and advertisement modules that enable the information to reach the users that will be characterised by interest in the given content. The audience will be carefully selected so that already at the start of dissemination the information enjoys positive opinions. This will in the long-term contribute to the discussion and the information will be discussed by the users for a longer time.

In the process of diffusion of innovation, considering the features of the audience, five groups are distinguished[16]:

- innovators
- early users
- early majority
- late majority
- marauders.

Information in social media reaches the people interested in the given group, person, profile or convictions the quickest. The media enable users to be notified if new information is made available. This refers to subscribing or following a

14 P. Kotler, *Marketing*, Ed. Dom wydawniczy REBIS Sp. z. o.o., Poznań 2005, p. 378.

15 M. Maurus, W. Zabłocki, *Zastosowanie teorii dyfuzji innowacji na przykładzie wprowadzenie na rynek Airbusa* A380, "Praca Naukowe Politechniki Warszawskiej", 2013, vol. 89, p. 137.

16 P. Kotler, *Marketing*, Ed. Dom wydawniczy REBIS Sp. z. o.o., Poznań 2005, pp. 273 and 281.

channel, person or profile. Therefore, the first recipients of information will be the persons sympathising with or supporting the perspective of their authority figure (early users). The next group is the early majority or friends of friends, the persons, to whom the information will be displayed due to the algorithm of the given social network or who will receive information due to its selection. The late majority is another group. They have many common features with the early majority, but they are sceptical. They consider the information as appropriate and use it following verification. The final group are users, who disagree with the information and have different opinions, thus criticising it and trying to demonstrate its flaws.

The innovators, or the group of people introducing the new innovation (information) on the market, are responsible for reaching the maximum audience. In case of information, they are also responsible for the content. However, in the media, sensation is more important than the reliability of information. The information itself, from creation to dissemination, undergoes a certain process – the information lifecycle. It is divided into five components[17]:

- creation – the users create content or obtain it from other sources
- management – content control by access rights management
- integration – opportunity to gain access from different system levels
- transaction – sales of contents, for example, using websites, to obtain funds
- distribution – dissemination of contents to the users.

Noticeably, the last stage of the information lifecycle is the dissemination of information to the audience. Consequently, it should be asked if the dissemination of false information (disinformation) needs to be carefully prepared so that the audience consider it as confident and reliable? Of course, this is a rhetorical question, because disinformation is intentional and false information.

Meaning of the Media for the Dissemination of Disinformation

The impact of the media world, in particular electronic media, on society in terms of formation of the opinion on the given phenomenon is increasing every year. Every year, more and more people become users of social media, so more and more people use the Internet. Thus, they move their activities to virtual worlds. This results in the increasing significance of social media in business, as

17 S. McKeever, *Understanding web content management systems: evolution, lifecycle and market*, Industrial Management & Data Systems 2003, 103(9), pp. 686–692.

they increase opportunities for success. The situation is similar in regard to the dissemination of disinformation. The impact of social media on society will be demonstrated here with myths and disinformation regarding vaccination.

The years 2020–2021 were recorded in history as the years of the pandemic of a virus similar to influenza, yet with a far greater lethality, COVID-19. The media generated much false information regarding vaccination, tests and cures for the virus. Most of it was based on earlier disinformation, which focused mainly on the illnesses that vaccines were said to cause.

The myths related to vaccination include the following[18]:

- the propagation of lifestyle diseases is caused by vaccination
- vaccines cause autism
- there are countries in which vaccines are not mandatory, so vaccines are not necessary in Poland
- vaccines are harmful for premature infants
- little children are overloaded with vaccines
- the anti-vaccine arguments are based on in-depth knowledge.

The first myth is the propagation of lifestyle diseases caused by vaccination. Undoubtedly, the development of medicine related to fighting diseases has reached the level, which has caused a drop in the lethality of infectious diseases. Unfortunately, this development was accompanied by an increase the susceptibility to non-infectious diseases, such as diabetes, obesity, allergies, cancers and autism. The Internet played a large role in the development of these diseases. Even now, many people search for diseases by symptoms – cancer is returned, even though the actual disease is harmless. In addition, in the virtual world we may find a lot of information on how vaccines spread lifestyle diseases. The problem in this argumentation is the correlation with vaccination, however extensive research proves that there is no connection between lifestyle diseases and vaccines.[19]

Another piece of disinformation in the media, very popular recently, is that vaccines cause autism. Even today, many people believe that vaccines cause autism in children. An example of false research related to autism is the notorious article of 1998, which was published in *The Lancet*.[20] The author described

---

18 https://poprostunauka.cm-uj.krakow.pl/6-mitow-na-temat-szczepionek/ [22.09.2022].

19 Morgan E., Halliday S.R., Campbell G.R., Cardwell C.R., *Vaccinations and childhood type 1 diabetes mellitus: a meta-analysis of observational studies*, Patterson C.C. 2016. Diabetologia, 59(2), pp. 237–243.

20 Wakefield A. J. et al. *Ileal-lymphoid-nodular hyperplasia, non-specific colitis, and pervasive developmental disorder in children*, The Lancet 1998, 351(9103), pp. 637-641.

the correlation between three-component vaccines and the occurrences of autism. Such vaccines are used against measles, mumps and rubella. Six years later, the investigation proved the results had been falsified. As a result, in 2010 the article was withdrawn and the author lost their professional licence. The evidence provided by the author of the false article was used for anti-vaccination activities to influence society.[21]

The third myth is about the need for vaccines in Poland. During the pandemic, the issue of mandatory vaccines was often raised. The purpose of the basic vaccines is to build herd immunity against a given disease. One Internet user produced a short video on how herd immunity works. The video shows dots in three colours:

- black – infected people
- blue – non-vaccinated people
- yellow – vaccinated people.

It is not difficult to guess that the more the vaccinated people, the lower the probability of infection.[22] Therefore, the vaccination mandate is necessary, because the immunity of the society increases.

The last myth related to the vaccinations is the knowledge of the anti-vaccination activists, namely their arguments. The entities undertaking such activities show isolated cases to demonstrate the negative impact and cause strong emotions. The knowledge presented to society usually comes from self-taught amateurs, not from professors of medicine. A noticeable phenomenon is the Dunning-Kruger effect, which describes the situation, in which unqualified people overestimate their own knowledge, while qualified people approach the subject with caution.[23] A similar effect can be seen in case of transport. Many people are afraid to take a plane, considering it dangerous, while they have no objections against using motor vehicles. Many statistics show that an airplane is the safest means of transport. The agitation of the human brain and presentation of a negative effect building emotions transforms into cognitive bias.

The use of the media to disseminate disinformation is also noticeable in statistics. For the purposes of the article, to show the disinformation in the media,

---

21 https://poprostunauka.cm-uj.krakow.pl/6-mitow-na-temat-szczepionek/ [22.09.2022].

22 Video available at: http://izp.wnz.cm.uj.edu.pl/cm/uploads/2018/11/rr4bz7xerbtjbdusoktw.mp4?_=1 [22.09.2022].

23 M. Popek, *Efekt Krugera – Dunninga*, article available at: http://www.acdesign.com.pl/wgrane-pliki/7-michal-popek_efekt-krugera-dunninga.pdf [22.09.2022].

information was sourced from demagog.org. For the purposes of study, the following was used:

- 2020 – 257 pieces of false information
- 2021 – 419 pieces of false information.

The number of pieces of disinformation shown above was detected by the demagog.org website. From the presented data, the authors collected information on user interactions with the posts published on the Facebook social media website. The posts on false information were collected at random, namely for every piece of disinformation spread, one post reproducing it was selected. The authors collecting data considered the following:

- post creation time
- disinformation (post) detection time
- number of reactions
- number of comments
- number of shares
- subject of the post.

The components shown illustrate the occurrence of disinformation in information space. They show the information lifecycle, because the more time passes on social media, the lower the interest. This is similar to the diffusion of innovation. The creation of a new sensational information may cause a drop in interest in the previous. This mechanism is used in the media to distract the public opinion.

Tab. 1. Subjects of disinformation – numbers in years 2020–2021

Subject	**Quantity**
education	**3**
other	**28**
medicine	**321**
COVID-19	106
face masks	18
medicine	31
vaccine	148
tests	18
politics	**95**
religion	**36**

(*continued on next page*)

Tab. 1. Continued

Subject	Quantity
technology	**37**
military	**4**
events in Poland	**68**
world events	**84**
Total:	**676**

Source: original work based on: https://demagog.org.pl/fake_news/ [22.09.2022].

Table 1 shows the quantity of false information by subject. Unsurprisingly, the highest quantity refers to disinformation related to medicine, as the most popular subject in 2020–2021 was the pandemic. The authors deliberately divided medicine into sub-topics:

- Covid-19
- face masks
- medicine
- vaccines
- tests.

The purpose was to demonstrate how the previously described myths were used to influence society. The greatest quantity of disinformation was related to vaccines, followed by the pandemic, medicine, face masks and tests. The most common topic related to vaccines was the debate on their effectiveness and causation of diseases. These topics enabled diverting the attention of public opinion.

The interest of the social media users is presented in Tab. 2.

Tab. 2. User interest – number of reactions, shares, comments

Subject	number of reactions	number of shares	number of comments
medicine	**207,261**	**495,592**	**33,149**
Covid – 19	72,234	243,644	9,860
vaccine	89,409	140,178	15,689
medicine	28,042	52,484	3,538
tests	9,280	38,843	2,089
face masks	8,296	20,443	1,973
politics	**49,936**	**107,717**	**13,372**
world events	**74,522**	**99,696**	**8,749**
other	**50,171**	**90,037**	**1,629**

Tab. 2. Continued

Subject	number of reactions	number of shares	number of comments
events in Poland	**39,442**	**77,150**	**35,896**
religion	**14,663**	**31,559**	**6,290**
technology	**20,082**	**20,329**	**2,780**
military	**5,263**	**532**	**2,202**
education	**865**	**224**	**464**
TOTAL:	**462,205**	**922,836**	**104,531**

Source: own study based on: https://demagog.org.pl/fake_news/ [22.09.2022].

The information presented is the total number of reactions, shares and comments of one post publishing the given information. This means that 676 posts were used for the study. The appropriate values were entered in the table and then summed up. The number of shares should be noted, because it indicates that a post is reproduced on another account or group. This number is the highest and over two years it is almost one million (922,836) shares of false information for 676 posts. This is surprising, because before undertaking the study, the authors expected a greater number of reactions. It should be noted that reproducing a given post results in another count of comments and reactions.

In the studies, for 676 posts, the number of reactions was 462,205, while the number of shares was 104,531. Thus, per one post with false information in 2020–2021, the average numbers were 684 reactions and 155 comments. Considering the shares and the average numbers, the capabilities of social media are apparent. Of course, the average values for the studied posts may be greater.

Another value to consider is the false information detection time. The authors present the detection time as the difference between the post-publication and the verification thereof by the demagog.org website.

Tab. 3. Average disinformation detection time in hours

Subject	Number of posts	Average detection time (in hours)
medicine	**321**	**354.77**
vaccine	148	358.72
Covid-19	106	277.86
medicine	31	481.73
tests	18	489.33

(*continued on next page*)

Tab. 3. Continued

Subject	Number of posts	Average detection time (in hours)
face masks	18	428.80
politics	**95**	**392.82**
world events	**84**	**625.23**
events in Poland	**68**	**186.72**
technology	**37**	**641.52**
religion	**36**	**1,041.18**
other	**28**	**633.60**
military	**4**	**114.00**
education	**3**	**24.00**
Total:	**676**	**437.92**

Source: original work based on: https://demagog.org.pl/fake_news/ [22.09.2022].

Tab. 3 shows that the average time to detect false information in 2020–2021 was 437.92 hours. The highest values are achieved by false information related to religion, then other, then world events. In case of disinformation related to medicine, the average value was 354.77 hours, which corresponds to 14.78 days.

In 2021, a study was conducted on the place of occurrence of false information. It was completed by GfK (Growth from Knowledge). The study was divided into[24]:

- traditional media
- social media
- interpersonal interactions
- communication applications
- other.

24 Poland; GfK; October 1–6, 2021; 817 respondents; 18 years and older; Individuals who have encountered fake news; Computer-assisted web interviews.

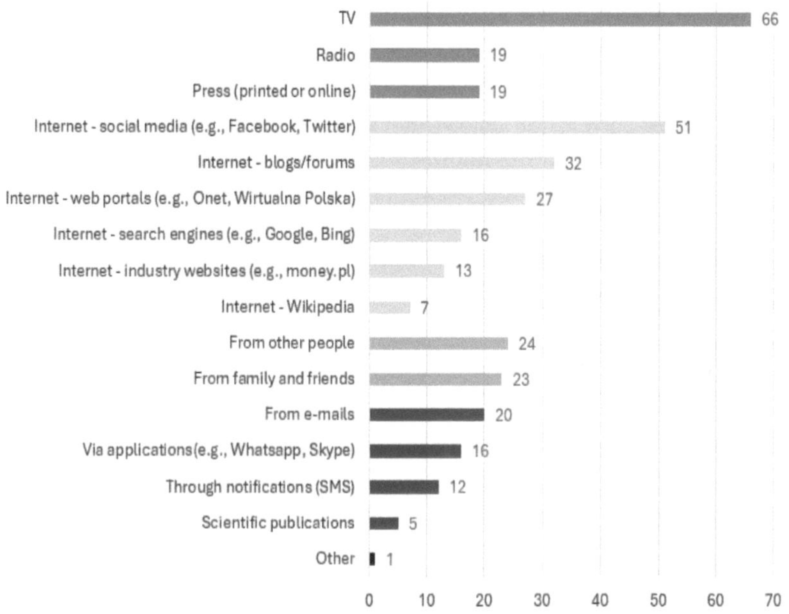

Chart 1. Place of occurrence of false information (in %)

Source: based on: https://www-1statista-1com-1zexc0dfi0051.han.wat.edu.pl/statistics/1319353/poland-fake-news-sources/ [02.10.2022].

The above data shows that two-thirds of the respondents indicate television as the source of false information. It should be noted that the study was conducted in 2021. The channel people use to find information may have changed. In Poland, much has been heard about the politicisation of television stations in order to reproduce information advantageous for the government. It is also worth noting the position of the social media indicated by every other respondent.

Speaking of disinformation in social media, the main causes of its production should be considered. According to the GfK study, the main reasons include[25]:

- intentional action – 75 %
- result of copying non-verified information – 37 %
- journalist error – 21 %

25 https://www-1statista-1com-1zexc0dfi0051.han.wat.edu.pl/statistics/1319356/poland-reasons-for-fake-news-emergence/ [02.10.2022].

- set of circumstances – 14 %
- difficult to say – 10 %

The most frequently indicated response is intentional action, which is not surprising, because such action follows the definition of disinformation. The next indicated response is copying information. At this point, one should return to the data on reproducing information, sharing (Tab. 2), which to some extent reflects this action.

Information, for example, on the Internet may be perceived as credible, supported by certain criteria.

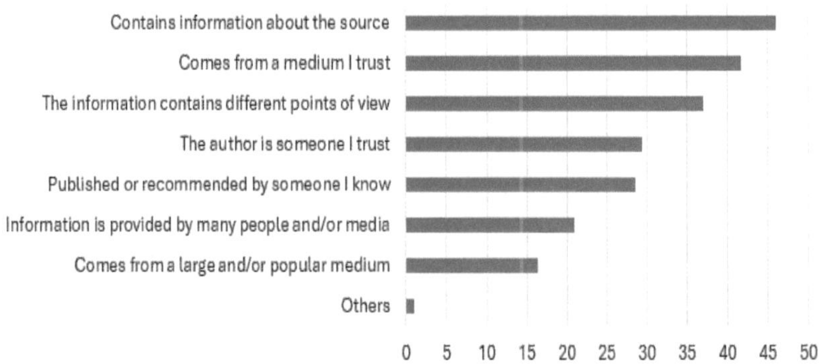

Chart 2. Most important criteria of credibility of information found on the Internet – Poland 2019

Source: based on: https://www-1statista-1com-1zexc0dfi00fa.han.wat.edu.pl/statistics/1009520/pol and-trust-in-information-found-on-the-internet/ [02.10.2022].

According to the respondents, the most credible information includes the source. This is the method of verification if the given piece of information is true. The following responses refer to authority and point of view. Unfortunately, credibility based on the *ad auctoritate* (appeal to authority) argument is not always true.

Conclusion

With technological development, social media have developed dynamically to become an inseparable part of daily life. Information in the media may take different forms, from true information to disinformation. Scientists differentiate thirteen types of social media. User interactions in social media may take four forms:

- publishing
- reacting
- commenting
- sharing.

These forms enable the users to act in different roles. The greatest responsibility in the dissemination of information is borne by the creators, which may be an organisation, group, association or a common user.

Social media will continue to gain in popularity, thus reducing the audiences of traditional media. The differences between these media are significant, in favour of social media. In particular, due to the ease of sharing information. One of the greatest advantages of social media is the capability of selecting the target audience for a given piece of information. Selection is possible thanks to advertisement and analysis modules.

The dissemination of information may be presented using the process of the diffusion of innovation. This process presents the subsequent audiences of information. The diffusion of innovation presents the final component of the lifecycle, that is, the dissemination of information. The information published both on social media and in traditional media first reaches the sympathisers, then the early majority and late majority, followed by critics and marauders in the end. The last users may extend the information lifecycle by reproducing it, while highlighting its weaknesses.

The authors presented the meaning of social media for the dissemination of information using the example of vaccinations. The article presented the myths related to vaccines. The main argument was that vaccines cause diseases. The wording of this argument may be presented as follows:

Vaccines are dangerous, because they cause autism.
Autism occurs
Therefore: vaccines are dangerous

There are studies that present it, however the problem in the argumentation is the correlation of the vaccines with lifestyle diseases.[26]

In the process of dissemination of disinformation, the media are very important. This is clearly shown by the statistics for 2020–2021, in which the numbers

26 Morgan E., Halliday S.R., Campbell G.R., Cardwell C.R., *Vaccinations and childhood type 1 diabetes mellitus: a meta-analysis of observational studies*, Patterson C.C. 2016. Diabetologia, 59(2), pp. 237–243.

of shares, reactions and comments are high. It should be mentioned that the statistics do not include the impressions of the given post.

To conclude, social media significantly contribute to the dissemination of disinformation, the main purpose of which is to gain approval of the users.

References

[1] Bączek P., *Dezinformacja jako zagrożenie dla bezpieczeństwa informacyjnego współczesnych społeczeństw*, [in:] Security Review 2020 no. 4(17).

[2] DiNucci D., *Fragmented Future*, „Print" 1999, 4 (53).

[3] Floridi L., Philosophical Conceptions of Information, [in:] Formal Theories of Information, Sommaruga G., Wyd. Springer, 2009.

[4] Hadnagy Ch., *Socjotechnika. Sztuka zdobywania władzy nad umysłem*, Wyd. Onepress 2020.

[5] Jachyra D., *Trollowanie – antyspołeczne zachowania w Internecie, sposoby wykrywania i obrony*, [in:] Zeszyty Naukowe Uniwersytetu Szczecińskiego, no. 656 (28) 2011.

[6] Kotler P., *Marketing*, Wyd. Dom wydawniczy REBIS Sp. z. o.o., Poznań 2005.

[7] Maurus M., Zabłocki W., *Zastosowanie teorii dyfuzji innowacji na przykładzie wprowadzenie na rynek Airbusa A380*, „Praca Naukowe Politechniki Warszawskiej", 2013, z. 89.

[8] McKeever S., *Understanding web content management systems: evolution, lifecycle and market*, Industrial management & data systems 2003, 103(9).

[9] Morgan E., Halliday S.R., Campbell G.R., Cardwell C.R., *Vaccinations and childhood type 1 diabetes mellitus: a meta-analysis of observational studies*, Patterson C.C. 2016. Diabetologia, 59(2).

[10] Morgan E., Halliday S.R., Campbell G.R., Cardwell C.R., *Vaccinations and childhood type 1 diabetes mellitus: a meta-analysis of observational studies*, Patterson C.C. 2016. Diabetologia, 59(2).

[11] Polczyk R., Szpitalak M., *Efekt wzmocnionej afirmacji – Wzrost odporności na dezinformację wskutek autoafirmacji wzmocnionej pozytywną informacją zwrotną*, Studia Psychologiczne t. 50 (2012), z. 3.

[12] Rogers E. M., *Diffusion of Innovations*, Wyd. Free Press, New York 2003.

[13] Skała M., *Manipulacja odczarowana! 777 skutecznych technik wpływu*, Wyd. HELION, Gliwice 2015.

[14] Wakefield A. J. et al. Ileal-lymphoid-nodular hyperplasia, non-specific colitis, and pervasive developmental disorder in children, The Lancet 1998, 351(9103).

[15] Aichner T., Jacob F. H., *Measuring the Degree of Corporate Social Media Use,* Available online: https://www.researchgate.net/publication/283073224_Measuring_the_Degree_of_Corporate_Social_Media_Use.

[16] Brzeski R., *Dezinformacja*, available online: https://socjocybernetyka.files.wordpress.com/2012/02/dezinformacja-skrypt-1.pdf [15.06.2022].

[17] Fleming G., Global Social Technographics Update 2011: US And EU Mature, Emerging Markets Show Lots of Activity, https://www.forrester.com/blogs/12-01-04-global_social_technographics_update_2011_us_and_eu_mature_emerging_markets_show_lots_of_activity/.

[18] Poland; GfK; October 1–6, 2021; 817 respondents; 18 years and older; Individuals who have encountered fake news; Computer-assisted web interviews.

[19] Popek M., *Efekt Krugera – Dunninga*, available online: http://www.acdesign.com.pl.

[20] http://izp.wnz.cm.uj.edu.pl/cm/uploads/2018/11/rr4bz7xerbtjbdusoktw.mp4?_=1.

[21] https://poprostunauka.cm-uj.krakow.pl/6-mitow-na-temat-szczepionek/.

[22] https://www-1statista-1com-1zexc0dfi0051.han.wat.edu.pl/statistics/1319356/poland-reasons-for-fake-news-emergence/.

Studies in Politics, Security and Society

Edited by Stanisław Sulowski

Vol. 1 Robert Wiszniowski (ed.): Challenges to Representative Democracy. A European Perspective. 2015.

Vol. 2 Jarosław Szymanek: Theory of Political Representation. 2015.

Vol. 3 Alojzy Z. Nowak (ed.): Global Financial Turbulence in the Euro Area. Polish Perspective. 2015.

Vol. 4 Jolanta Itrich-Drabarek: The Civil Service in Poland. Theory and Experience. 2015.

Vol. 5 Agnieszka Rothert: Power of Imagination. Education, Innovations and Democracy. 2016.

Vol. 6 Zbysław Dobrowolski: Combating Corruption and Other Organizational Pathologies. 2017.

Vol. 7 Vito Breda: The Objectivity of Judicial Decisions. A Comparative Analysis of Nine Jurisdictions. 2017.

Vol. 8 Anna Sroka: Accountability and democracy in Poland and Spain. 2017.

Vol. 9 Anna Sroka / Fanny Castro-Rial Garrone / Rubén Darío Torres Kumbrián (eds.): Radicalism and Terrorism in the 21st Century. Implications for Security. 2017.

Vol. 10 Filip Pierzchalski: Political Leadership in Morphogenetic Perspective. 2017.

Vol. 11 Alina Petra Marinescu: The Discursive Dimension of Employee Engagement and Disengagement. Accounts of keeping and leaving jobs in present-day Bucharest organizations. 2017.

Vol. 12 Jacek Giedrojć: Competition, Coordination, Social Order. Responsible Business, Civil Society, and Government in an Open Society. 2017.

Vol. 13 Filip Ilkowski: Capitalist Imperialism in Contemporary Theoretical Frameworks. 2017.

Vol. 14 Leszek Leszczyński / Adam Szot (eds.): Discretionary Power of Public Administration. Its Scope and Control. 2017.

Vol. 15 Tadeusz Klementewicz: Understanding Politics. Theory, Procedures, Narratives. 2017.

Vol. 16 Tomasz Bichta: Political Systems of the Former Yugoslavia. 2018

Vol. 17 Miroslav Palárik / Alena Mikulášová / Martin Hetényi / Róbert Arpáš: The City and Region Against the Backdrop of Totalitarianism. 2018

Vol. 18 Jolanta Itrich-Drabarek / Stanisław Mazur / Justyna Wiśniewska-Grzelak (eds.): The Transformations of the Civil Service in Poland in Comparison with International Experience. 2018..

Vol. 19 Jerzy Juchnowski / R. Jan Sielezin / Ewa Maj: The Image of "White" and "Red" Russia in the Polish Political Thought of the 19th and 20th Century. Analogies and Parallels. 2018.

Vol. 20 Roman Kuźniar: Europe in the International Order. 2018.

Vol. 21 Piotr Jaroszyński: Europe - the Clash of Civilisations. 2018.

Vol. 22 Stanisław Filipowicz: Truth and the Will to Illusion. 2018.

Vol. 23 Andrzej Szeptycki: Contemporary Relations between Poland and Ukraine. The "Strategic Partnership" and the Limits Thereof. 2019.

Vol. 24 Sylwester Gardocki / Rafał Ożarowski / Rafał Ulatowski (eds.): The Islamic World in International Relations. 2019.

Vol. 25 Jacek Zaleśny (ed.): Constitutional Courts in Post-Soviet States. Between the Model of a State of Law and Its Local Application. 2019.

Vol. 26 Andrzej Antoszewski / Przemysław Żukiewicz / Mateusz Zieliński / Katarzyna. Domagała: Formation of Government Coalition in Westminster Democracies. Towards a Network Approach. 2020.

Vol. 27 Joanna Osiejewicz: Global Governance of Oil and Gas Resources in the International Legal Perspective. 2020.

Vol. 28 Anita Oberda-Monkiewicz: Poland-Mexico towards a Strategic Partnership. 2020.

Vol. 29 Bartosz Czepil / Wojciech Opioła: Ethnic diversity and local governance quality. The case of Opole Province in Poland. 2020.

Vol. 30 Adam Szymański / Jakub Wódka / Wojciech Ufel / Amanda Dziubińska: Between Fair and Rigged. Elections as a Key Determinant of the 'Borderline Political Regime' - Turkey in Comparative Perspective. 2020.

Vol. 31 Tomasz Grzegorz Grosse (ed.): Fuel for Dominance. On the Economic Bases of Geopolitical Supremacy. 2020.

Vol. 32 Dariusz Jarosz / Maria Pasztor: From Subjection to Independence. Post-World War II Polish-Italian Relations. 2020.

Vol. 33 Paweł Sekuła: Chernobyl Liquidators. The Unknown Story. With the Testimony of the President of Latvia. 2020.

Vol. 34 Kamil Glinka (ed.): Urban Policy System in Strategic Perspective: From V4 to Ukraine. 2020.

Vol. 35 Paweł Lesiński: At the Origins of German Liberalism: the State in the Thought of Robert von Mohl. 2020.

Vol. 36 Jana Popovicsová (ed.): Reflexions about a Cultural and Social Phenomenon: Identity. 2020.

Vol. 37 Agnes Bernek: Geopolitics of Central and Eastern Europe in the 21st Century. From the Buffer Zone to the Gateway Zone. 2021.

Vol. 38 Marek Antoni Musioł: The European Union as a Post-Lisbon Regional Security Complex. 2021.

Vol. 39 Alina Petra Marinescu: Manipulation in the Disclosure of the Securitate Files. The Case of Mona Muscă. 2021

Vol. 40 Jiří Zákravský: Cycling Diplomacy. Undemocratic Regimes and ProfessionalRoad Cycling Teams Sponsorship. 2021.

Vol. 41 Mirosław Karwat / Filip Pierzchalski / Marcin Tobiasz (eds.): Constituents of Political Theory. Selected Articles of the Warsaw School of Political Theory. 2021.

Vol. 42 Angieszka Bieńczyk-Missala: Preventing Mass Human-Rights Violations and Atrocity Crimes. 2021.

Vol. 43 Jan Garlicki: Political Participation Capital. 2021.

Vol. 44 Mirosław Karwat: Theory of Provocation. In Light of Political Science. 2022.

Vol. 45 Anna Sroka / Piotr Potejko / Rubén Darío Torres Kumbrián (eds.): Migration and Border Security. Global Perspectives. 2021.

Vol. 46 Stanisław Sulowski / Tomasz Słomka (eds.): The Political System of Poland. Tradition and Contemporaneity. 2022.

Vol. 47 Ewelina Kancik-Kołtun / Josef Smolík (eds.): New Political Parties in the Party Systems of the Czech Republic. 2022.

Vol. 48 Ewa Feder-Sempach: Economic Populism in British and American Political Discourse. A Comparative Analysis of Boris Johnson's and Donald Trump's Speeches. 2022.

Vol. 49 Katarzyna Kołodziejczyk (ed.): The EU Towards the Global South During the COVID-19 Pandemic. 2022.

Vol. 50 Silvia Amato: Public Undertakings of Nuclear Waste Storage. The Role of the Government in the Dissemination of Public Knowledge. 2022.

Vol. 51 Agnieszka Cholewińska-Nielsen: Evolving Humanitarianism. The Work of the International Red Cross and Red Crescent Movement. 2022.

Vol. 52 Josef Smolik / Ewelina Kancik-Kołtun: New Political Parties in the Party System of Poland. 2023.

Vol. 53 Rafał Wiśniewski / Elżbieta Szyszlak / Radosław Zenderowski (eds.): Cultural Security. Theory – Selected Aspects – Case Studies. 2023.

Vol. 54 Filip Pierzchalski: Political Leadership: Structure – Consciousness – Emotions. 2023.

Vol. 55 Łukasz Zamęcki / Li-Jiuan Chen-Rabich (eds.): The Democratisation Process of Poland and Taiwan: Comparative Perspective. 2023.

Vol. 56 Marcin Gajek: Refusing to be Forgotten: Southern Conservatism and the Political Thought of M. E. Bradford. 2023.

Vol. 57 Sulowski Stanisław (ed.): Security Challenges at the Dawn of a New International Order. 2023.

Vol. 58 Alicja Z. Nowak / Kinga Anna Gajda (eds.): University and War in Ukraine. 2023.

Vol. 59 Bernard Wiśniewski / Paulina Polko (eds.): Contemporary Understanding of Security and Its Contexts. 2024.

Vol. 60 Aleksander Żołnierski / Dariusz Jaruga (eds.): Big Data. Post-truth and Disinformation in Politics and Economy. 2025.

www.peterlang.com

www.ingramcontent.com/pod-product-compliance
Ingram Content Group UK Ltd.
Pitfield, Milton Keynes, MK11 3LW, UK
UKHW041826031026
15587UKWH00002B/591

9 783631 92243